A   **DATE DUE**

A Measure of Everything

An illustrated guide to the science of measurement

General Editor: **Christopher Joseph**

Contributors: **Adam Parfitt, David Price and Marcus Weeks**

FIREFLY BOOKS

A Firefly Book

Published by Firefly Books Ltd. 2005

Copyright © 2005 Ivy Press Limited

First printing

Published in the United States by
Firefly Books (U.S.) Inc.
P.O. Box 1338, Ellicott Station
Buffalo, New York 14205

Published in Canada by
Firefly Books Ltd.
66 Leek Crescent
Richmond Hill, Ontario L4B 1H1

This book was conceived, designed and produced by
THE IVY PRESS LIMITED
The Old Candlemakers, West Street,
Lewes, East Sussex BN7 2NZ, U.K.

Illustrator – Ivan Hissey
Consultant – Nigel Hawkes

Creative Director – Peter Bridgewater
Publisher – Sophie Collins
Editorial Director – Jason Hook
Project Manager – Stephanie Horner
Art Director – Karl Shanahan
Designer – Ginny Zeal

Publisher Cataloging-in-Publication Data (U.S.)

A measure of everything : an illustrated guide to the science of measurement
Christopher Joseph [editor]. [224] p. : cm.
Includes index.
Summary: A wide-ranging guide to units of measure including earth and life sciences, physical sciences and technology, and leisure. Topics include astronomy, medicine, temperature, chemistry, food, currency, and photography.
ISBN 1-55407-089-9 (pbk.)
1. Mensuration. 2. Weights and measures.
3. Physical measurements.
I. Joseph, Christopher. II. Title.
530.8 dc22   QA465.M437 2005

Library and Archives Canada.
Cataloguing in Publication

A measure of everything : an illustrated guide to the science of measurement / general editor, Christopher Joseph : contributors, Adam Parfitt, David Price, Marcus Weeks.
Includes index.
ISBN 1-55407-089-9
1. Weights and measures-
-Handbooks, manuals, etc.
I. Joseph, Christopher II. Parfitt, Adam, 1972- III. Price, David, 1955-
IV. Weeks, Marcus V. Title.
QC90.5.M42 2005   530.8'1
C2005-901500-4

Printed in China

Contents

Introduction

Measurement, in one form or another, is one of mankind's oldest and most vital activities. Even before the dawn of civilization, relative measurement—"their tribe is bigger than ours"—was vital to the survival of any individual or group. The members of a hunter-gatherer society needed the concepts of "more," "less" and "enough" (enough time to get home before dark; enough food to ensure that no one will have to go hungry). In fact, when you think about it, the human ability to determine whether the five people gathering berries have gathered enough for the entire tribe is a remarkably sophisticated piece of calculation.

As time passed, and human civilization evolved, such purely relative measurements were no longer always sufficient. With the creation of permanent settlements of ever-increasing size such estimation was, in any case, rather more difficult. The increasing sophistication of language allowed comparisons to become ever more complex—enough for one person is not always enough for another.

The earliest historical record of a unit of measurement is the Egyptian cubit, in around 3000 B.C.E., decreed to be equal to the length of a forearm and hand (from the back of the elbow to the extended tip of the middle finger) plus the width of Pharaoh's palm.

The pyramids at Giza in Egypt are geometric constructions of astonishing accuracy, even if the symbolic meaning of their proportions is not fully known. The largest of the three, Khufu's pyramid, remained the tallest manmade structure on Earth from its construction before 2500 B.C.E. until the 19th century—more than 4,300 years later.

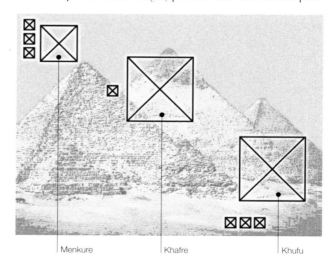

Menkure   Khafre   Khufu

This is, of course, still rather flexible—my forearm is not the same length as yours, and since neither of us met Pharaoh, we cannot confidently say precisely how wide his palm was. It is not yet much of a step forward from measuring the distance in hand-widths, paces or whatever other approximation springs to mind.

By 2500 B.C.E., the necessary leap forward had been taken. The complicated and rather imprecise definition had been simplified drastically: a cubit was the same length as the prototype cubit, stored safely from harm. This was the "royal master cubit," a black marble rod some 52 cm in length—and the size of the user's forearm was no longer an issue. From this simple model it becomes possible to measure many things: distances, areas, volumes, even masses. (This last was defined relative to the mass of a specified volume of a specified substance— usually water or gold, but many other things have been used.)

With accurate measurement, many things that had been difficult or impossible before were now achievable. Trade (between people, towns or regions) became more sophisticated once it was possible to exchange the precise same quantity of grain for the same quantity of gold (or salt, or fabric, or whatever else was needed) every time.

The most important change brought about by reliable measurement, however, was the possibility of science. Without accurate measurement—and equally accurate recording—no useful form of science, engineering or technology is really possible. Without accurate measurement, you couldn't be holding this book in your hands—a modern printing press is a huge piece of precision machinery. Dozens or hundreds of carefully designed and accurately made components move a sheet of paper (which must be a particular size) exactly the right distance, at exactly the right speed, for the precisely calculated amounts of ink (in four carefully defined colors) to fit together on the page as text and color diagrams—without disappearing off the edges.

In ancient Egypt, the stars rose and fell in the sky at night, as they had always done. Now, however, they were watched by careful, thoughtful men who measured their movements—the maximum height to which they rose above the horizon, the distance between the points at which they appeared on different nights, and so on. Probably their most important discovery (made by Egyptian astronomers watching the star Sirius) was that the time taken for Earth to orbit the sun was 365 and a quarter days. The Egyptian priesthood, unfortunately, was not impressed. The gods would not have been so foolish, and the priests insisted that the calendar should remain fixed at 365 days exactly. This may have seemed more sensible to them, but it forced their official calendar (with dates precisely calculated for sowing seed, harvesting crops and religious festivals of all kinds) to shift by six hours every year relative to the actual changes of the seasons.

Elsewhere, other civilizations also measured the skies—without, unfortunately, leaving us the detailed records that the Egyptians kept. Their awe-inspiring creations, of which Stonehenge is perhaps the most famous, leave us with no doubts about their designers' ability to measure—the sun shines precisely down the main avenue of

Although the famous bluestone slabs arrived at Stonehenge in the west of England from Wales in around 2500 B.C.E., the earliest similar construction on the site is believed to have been built some 600 years earlier.

summer solstice point 1749 B.C.E.

avenue/solstice line

Stonehenge at the summer solstice. But we have no idea what these immense constructions (as remarkable in their way as the Egyptian pyramids, or the Great Wall of China) were for.

Six hours in a year seems only a very small inaccuracy, but since the change was always in the same direction it accumulated over time, and, after some 730 years, the Egyptians found themselves celebrating midsummer on the shortest day of the year. The drift continued, through Egypt's long history, with the priests preferring their sacred calendar to the astronomers' detailed and painstaking measurements, until the Romans arrived. Julius Caesar listened to the star-gazers, and added the extra day every four years they recommended to the Roman calendar: his successor Augustus then finally forced the Egyptians to do likewise, bringing their year into line with the rest of the Empire.

The Romans were great engineers, and measurement was important to them in many ways, even if their assumptions were not always correct. Their influence can still be felt today, in an astonishing variety of ways, though the story that the standard gauge of modern railways is derived from the width of a Roman chariot is, alas, a myth. (Most vehicles pulled by two horses are around that width, but the gauge became "standard" because it was the one the best engines were built for in the greatest numbers—the standard width might as easily have been the 3 foot 6 inches still used in parts of Australia, or Brunel's faster and more stable 7-foot width.)

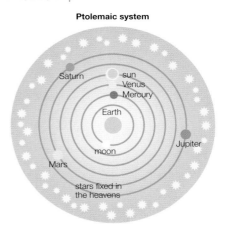

Ptolemaic system

We owe to the Romans many of the names of our traditional units—and their language has also been plundered for scientific terms. "Ounce" and "inch" both derive from the Roman *uncia*, meaning a twelfth. Uses of *uncia* were not limited merely to weight and distance, however. Almost anything could sensibly be divided into 12 parts, and almost anything (at one time or another) was. To a Roman it made perfect sense to talk of *unciae* of a cake, or an estate—perhaps even a business with ownership divided among several people.

For centuries after the fall of Rome, systems of customary units remained largely unchanged; there was, after all, nothing new under the sun to measure. Still, the feudal monarchs of Europe were interested in measurement for the same reason that any government is: if you do not know what a man owns or produces, how can you tax him on it?

As the Renaissance reinvigorated progress in art and science, interest in measurement regained some of its former vigor. Problems, of course, still occurred. Galileo, Copernicus and others suffered imprisonment (or worse) at the hands of the established Church, for revealing the conclusions they drew from their measurements, despite the fact that they were all deeply religious men, who marveled at what they saw as the revelation of the amazing sophistication and complexity of God's creation, and contradicted nothing in the Bible.

Traditional units of measurement continued in use all over Europe, sometimes more different from one town to the next than they were from one century to the next. Steadily, though, they became more accurate and reliable. Candles with hour markings on (where the candle might burn at different rates, or the markings be spaced slightly differently) were replaced by clockwork. While a clock in one town might tell a different time from that in a town 20 miles away, both could be relied on to make each hour the same length, and the hours of one would be the same as the hours of the other.

Then came the French Revolution, and the greatest reform since the introduction of the Gregorian calendar. The metric system, it was

In his astronomical treatise Almagest, *one of the most influential books of Antiquity, Ptolemy compiled the astronomical knowledge of the ancient Greek and Babylonian world. This geocentric model of the solar system remained the generally accepted model in the western and Arab worlds until it was superseded by the heliocentric solar system of Copernicus.*

9

declared. was "for all people. for all time." Napoleon soon followed and. being the man he was. intended to enforce this with military might wherever it was necessary (or possible) to do so. Though initially unpopular even in France. this system was the logical measuring accompaniment of the decimal numbering system that had by then firmly supplanted Roman numerals throughout Europe. Decimal numbers had made finances of all kinds (including tax collection) much more straightforward. as well as greatly simplifying much mathematical thinking. Trade and measurement. however. was another matter: 10 sheep were 10 sheep. but a pound was still 12 ounces. Conversions between different types of measurement were more complex still—if one pint of something weighed one pound. one gallon of it most definitely did not weigh one stone. The metric system supplanted all of this with a system of related units. with only one unit for any given type of measurement. distinctions of scale being made purely by the use of prefixes to indicate the magnitude of the number.

The standard unit forming the base of this system. from which all of the others were initially derived. was the meter. It was defined as one-ten-millionth of the distance from the Equator to the North Pole along a meridian passing through Paris (in a steady curve at a theoretical sea level. rather than along the less-than-smooth surface of Earth). An enormous survey was carried out at great expense. and over several years. to calculate this distance as accurately as possible. and a platinum-iridium bar—the equivalent of Pharaoh's master cubit—made as a permanent record. Carefully manufactured exact duplicates were distributed to standards authorities throughout Europe (and. ultimately. the world) to ensure that everyone was using the same meter.

Although many other SI units have names, they can all be defined in terms of the seven base units.

meter   kilogram   second   ampere kelvin   mole   candela

It is interesting to note that many of what are now thought of as the basic units of the metric system (or. as it subsequently became. the SI—Système Internationale) were not originally intended to be so. The original official base unit for mass was the gram. far too small to be any use to most people. Similarly. there was no official base unit for volume: lengths were measured in meters. so volumes were to be measured in cubic meters. Unfortunately. a cubic meter was far too large a volume for most people to use for the goods they bought and sold in the marketplace. The French may be famous for their wine. but

1 cubic meter is more than 1,300 bottles—at two glasses per day, it would last over a decade. Public demand led to the introduction of a smaller unit of volume: the liter. It was defined as 1 cubic decimeter, making the mass of 1 liter of water equal to 1 kilogram—in theory.

Unfortunately, a later repeat of the survey using newer and more accurate instruments gave a slightly different value to the meter. It was decided that the prototype was "good enough" and would constitute the official definition of the meter from that time on. Unfortunately (again) continuing improvement in scientific equipment, and ever more accurate measurements, revealed that not only did the platinum-iridium prototypes for the meter and the kilogram not match their intended definitions, but they might actually be changing as time passed. After some thought, the 19th-century committee decided to retain the prototypes—and therefore the current values of the units—rather than make a fresh attempt to match the definition. (Interestingly, while the meter has subsequently been firmly redefined to bypass the problem, the kilogram is still defined by the metal prototype. A number of possible alternatives based on the masses of subatomic particles or stable isotopes have been suggested, but none has yet been officially instituted.) This then meant that the mass of 1 liter of water was in fact by definition not a kilogram—though the difference is only a few hundredths of a gram, too small to be of any importance for most purposes.

By this time science and measurement had become inextricably entwined. While accurate measurement is a necessary prerequisite for good science, the continuing advancement of science through the 18th, 19th and 20th centuries provided an unending stream of new things to measure and new ways of measuring old things.

In the 1960s, the C.G.P.M. (Conférence Générale des Poids et Mesures), which meets every few years to discuss and resolve problems with the SI units and their use, decided enough was enough. It introduced sweeping redefinitions, with six base units, including the meter and kilogram, along with the kelvin, the second, the ampere and the candela. The useful chemical term mole became a seventh soon afterward. From these, everything else could be defined. The base units (except the kilogram) were carefully redefined in terms of reliably reproducible physical measurements, so that they could be tested and confirmed by anyone with the time, equipment and desire to do so.

Minor changes and tweaks were made to the system, but by 1968 most of the definitions had settled into the form they hold today. One of the last major changes was the redefinition of the meter in 1983, from a certain number of wavelengths of a particular type of electromagnetic radiation emitted under certain circumstances, to the distance that light travels through a vacuum in a specific fraction of a second. Great care was taken to retain the values (which had been

The complex and fragile nature of survey equipment would make the original definition of the meter an impressive achievement even today. Using the equipment available in the late 18th-century, including early forms of theodolite as pictured here, and traveling on rough roads in horse-drawn carts, it was remarkable indeed.

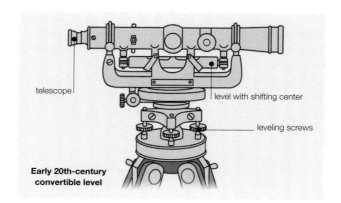

telescope

level with shifting center

leveling screws

**Early 20th-century
convertible level**

constant, by this time, for nearly 200 years) by the use of numerical "fudge factors." For this reason light travels 299,792,458 meters per second, rather than a nice round 300,000,000, which would make the math simpler, as well as being much easier to remember.

Great care was taken to ensure that the units of the SI were what is called "coherent." This is best explained as the consistency of scale between units. Thus, for example, the pascal is coherent with the newton and the meter because a pressure of 1 pascal is equal to a force of 1 newton acting on an area of 1 square meter.

Another advantage of the system is that, if you want to describe a thousand million tonnes (either one billion, or one milliard, depending on your chosen naming system for large numbers), it is not necessary to come up with a new word. You merely check the list of SI-approved prefixes and there you are: one gigatonne. Even so, by the late 20th century, as scientists discovered ever more about the universe, and wanted to measure ever larger—and ever smaller—things, the system needed expansion. Astronomers at one end of the scale and physicists, investigating subatomic particles, at the other were running out of prefixes. In the 1970s and 80s, therefore, the C.G.P.M. extended the list of approved prefixes at both ends of the scale.

Even those countries still using such traditional units and measures as the pint, the pound and the mile have long since redefined them in terms of their metric equivalents. Many other inventions of the 1790s have fallen by the wayside, but the early metrication slogan "for all people, for all time" seems increasingly appropriate.

Scientists, of course, continue to investigate new and different things, many of which are measured purely in terms of the standard SI units. Sometimes there is a logical, sensible unit (e.g., the light-year) which merely happens not to be a convenient power of ten, but these can still easily be stated in terms of the SI unit. Sometimes—as with the barn and shed in subatomic physics—the unit is a convenient power of 10, and people merely like the sound of the name better. The use of such terms is never officially approved by the SI, but they are usually quicker and easier, and have a friendlier feel than, say, "100 square femtometers." Humans like things that are familiar—that feel right. This may well help to explain why none of the attempts to create a metric clock has ever been particularly popular: it has no advantages, and it feels wrong. The fact that such designs usually result in a longer working week may also, of course, have some slight influence…

We know of many different units used by one group of people or another over the millennia, and likely there were many more for which we have no record. The vast majority of those for which we do have records have fallen into disuse. Some have long and detailed histories, explaining how (and by whom) they were defined, what they were used for, and how they changed as time passed. For others we have only the barest records—sometimes just a name, with no indication of what fixed value they possessed (if indeed they did—the amphora and the barrel have both been standard units of measurement, but both were also types of container, and could come in many different sizes).

The values of many others varied wildly with place or time. Many ancient units are known only to historians; others are more widely known, if not always correctly understood. The talent, for example, is referred to frequently in the Bible, and everyone knows it is a sum of money. But the value of a coin, in the days of the Old Testament, was normally just the value of the precious metal that it contained—and a talent was a great sum of money indeed: at 25 kg, rather heavier than any practical coin.

Science is intimately bound up with the story of measurement: each drives the other forward. But measurement, either conscious or instinctive, is a part of every human activity—choosing the right color, drawing in perspective, correctly valuing a house for sale, or fitting the correct number of syllables in a line of poetry.

This book is intended to provide a friendly guide to the world of measurement, but it is necessarily only a whistle-stop tour. A complete list of all the units of which we have records (or even those to which we can provide a definite value) would be as large as any dictionary—even without including the definitions of (and differences between) the properties those units measure, and the instruments that record them. In compiling such a book as this, many things must be excluded as too rare to be worth defining or too common to need it. Our goal has been to produce something that is both a useful reference book—when you need one—and an entertaining read—when you don't. I hope that we have succeeded, and that nothing has been left out by accident.

How the entries work

Within each entry, units of measure are highlighted in **bold type** where the reader may wish to cross-refer to another entry (via the index) for further detail or explanation. The symbol 🍧 indicates entries that are illustrated.

13

Talent weight

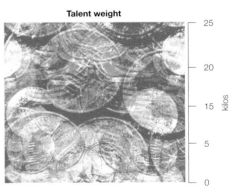

25

20

15   kilos

5

0

At 25 kg, a talent of gold is around U.S. $350,000 at 2005 prices, and would have made you a very rich man indeed in biblical times.

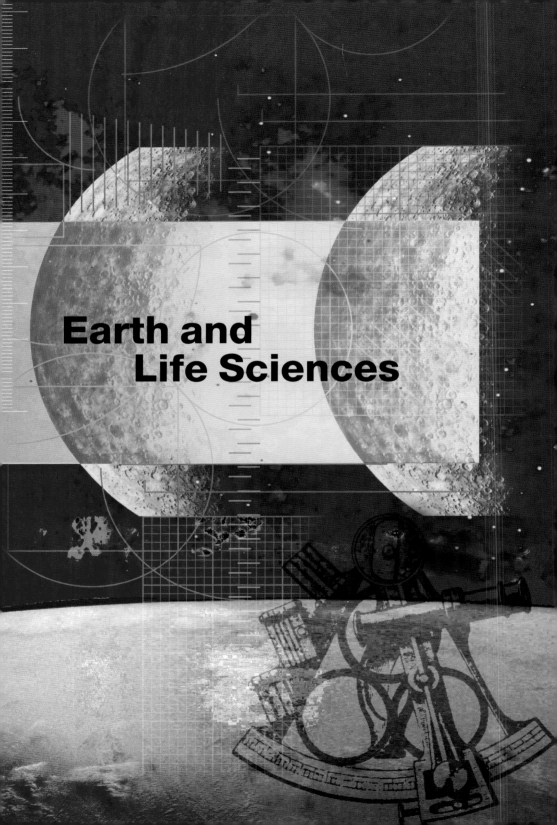

Earth and Life Sciences

astronomy and astrology

aberration

the difference between the actual position of a heavenly object and its apparent position as seen from Earth. This discrepancy is caused by relative sideways movement, an object's apparent position being slightly displaced (relative to its actual position) in the direction of the Earth's movement as Earth orbits the sun, and spins on its axis (though the latter effect is never more than a fraction of a degree).

eccentricity

the extent to which a curved path (such as the orbit of a planet or comet around the sun) differs from a perfect circle, measured as the difference between its long and short axes, divided by the sum of those axes. An object with a highly eccentric orbit will come very close to the sun at one end of the orbit, with the other end being much farther away, whereas circular or almost circular orbits have the sun at, or near, their center. For an object in actual orbit, eccentricity always has a value between zero (a perfect circle) and one. An eccentricity of greater than one indicates an open curve, rather than a closed loop.

apoapsis

the point in its orbit at which the orbiting object is farthest from whatever it is orbiting. For an object orbiting the sun, apoapsis is called aphelion. An object in Earth orbit is at apogee when it reaches its greatest distance from the planet. Each other planet in the solar system also has an individual name for its satellites' apoapsis. These are formed from the Greek names for the gods the planets are named after, and so are not always the same as for the planets themselves (which sometimes use the Roman name for the same god): e.g., apoapsis of a satellite of Mars is apoareion.

periapsis

the opposite of **apoapsis**: i.e., the point at which the distance between orbiting and orbited objects is smallest. For the sun's satellites, periapsis is called perihelion; for satellites in Earth orbit it is perigee. The moon's

As the eccentricity of a body's orbit increases, so does the difference between its apoapsis and periapsis.

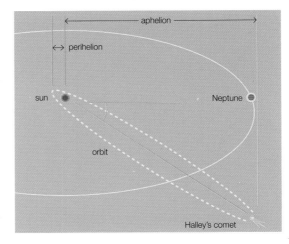

15

perigee is at around 359,000 km from Earth, some 42,000 km closer than its apogee. Halley's comet has its perihelion about 0.6 **AU** from the sun; the comet's aphelion is at 35.3 AU.

inclination

the angle between the orbital planes of two astronomical objects—usually, between an object and Earth. The orbital plane of an object is the flat surface that contains its entire orbit.

angular distance

the angle between two heavenly objects if they are projected as points onto a **celestial sphere** centered on Earth.

precession

the gradual change of direction of the axis of rotation of a rotating object (e.g., a planet, or spinning top) due to applied torque. In the case of celestial objects, precession is caused by gravitational tidal forces (those of the sun and moon, for Earth) attempting to realign the spinning planet, which is not quite spherical. Precession causes a slow change in the positions of the stars as measured in the equatorial **celestial coordinate** system. It also causes the equinoctial points to move westward along the ecliptic by around 5/6 of a second of arc each year, termed the precession of the equinoxes.

celestial sphere

an infinitely large imaginary sphere, centered on Earth. Every heavenly object (planets, stars, satellites, etc.) can be represented by a point on the celestial sphere, at the position where a line through the object from the center of Earth would touch the sphere.

Different kinds of celestial coordinates are used by astronomers for different purposes.

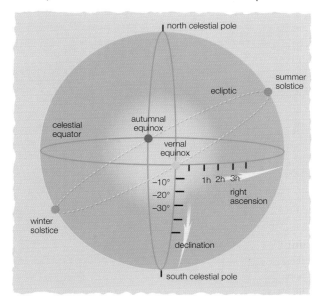

✥ celestial coordinates

any system for mapping the position of objects on the celestial
sphere. The main system used by professional astronomers is the
"equatorial" system, which uses **declination** and **right ascension**
to define the position of an object absolutely. Others, notably amateur astronomers and newspaper columnists, prefer the
"horizontal coordinate" system, which uses **azimuth** and **elevation** to define the position of the object relative to the observer's
location. The latter system is far easier to use instinctively, but the
numbers it produces are entirely local—an object at **zenith** (i.e.,
90°elevation) in Washington D.C., for example, will be much lower
in Los Angeles. There are many other celestial coordinate systems,
often designed for special purposes (e.g., one uses declination and
ascension relative to the plane of the Milky Way galaxy for measurement of the galaxy itself).

*Azimuth and altitude are
useful for finding positions
relative to the observer.*

ecliptic coordinates

a set of celestial coordinates widely
used by astrologers, as it is most
useful for measuring the position of
objects within the solar system.
The ecliptic is the apparent path
the sun follows around the celestial
sphere over the course of a year.
Ecliptic longitude is measured
around the path from the point of
the vernal (spring) equinox, while
ecliptic latitude is the distance
north or south from the ecliptic.

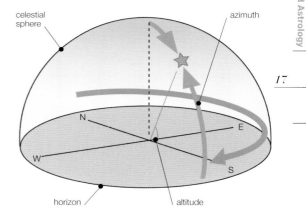

declination

the **angular distance** of a celestial object from the projection of
the Earth's Equator onto the celestial sphere. Positive declination
means the object is to the north of the Equator; objects with negative declination are to the south of it.

right ascension

the **angular distance** eastward from a vertical circle through the
vernal equinox to the point where an object appears on the celestial sphere. Right ascension is normally measured in hours (equal
to 15° of arc), minutes ($1/60$ of an hour) and seconds ($1/60$ of a
minute). The combination of declination and right ascension forms
the "equatorial" set of **celestial coordinates**.

✥ azimuth

the **angular distance** measured clockwise along the observer's
horizon from the north point to the line from zenith to nadir passing through the object's apparent point on the celestial sphere.

elevation

the **angular distance** of a celestial object above the observer's theoretical horizon (a horizontal circle around the observer projected onto the celestial sphere). Elevation is sometimes also called the altitude of the object.

depression

the angular distance of a celestial object below the observer's theoretical horizon (a horizontal circle around the observer projected onto the celestial sphere). Objects are occasionally described with a negative **elevation** value. rather than a depression. but never the other way round.

zenith

the point on the imaginary **celestial sphere** directly above the observer. That is. the point at which a line drawn from the center of Earth through the observer meets the celestial sphere. Sometimes also used (especially by astrologers) to mean the point of an object's path around the celestial sphere at which it is highest over the observer's horizon. but in astronomical terms this is an incorrect usage. The opposite point (directly beneath the observer's feet) is called the nadir.

elongation

the angular distance (in a direct line) between a celestial object and the sun. as seen from Earth. The inverse version (the angular distance between Earth and the sun. as seen from another celestial object) is called that object's phase angle.

orbital period

the length of time taken for an object to complete one orbit around whatever it may be orbiting: this is dependent on orbital velocity. eccentricity and its **apo-** and **peri- apses**. Earth's orbital period (around the sun) is one year. while that of Halley's comet is a little over 76 years. The moon's orbital period around the Earth is a lunar month.

orbital velocity (orbital speed)

the **speed** (not **velocity**) at which an object in orbit travels through space relative to the body being orbited. For a circular orbit. orbital velocity is easily calculated from the radius of the orbit and the orbital period. It can. however. also be calculated for elliptical orbits and even parabolic passing trajectories. because the combined kinetic and gravitational potential energies of the orbiting object will remain constant. The speed can therefore be calculated or measured for a specific instant. and the change in kinetic energy as the object moves calculating from the known change in potential energy.

✿ parallax

the apparent change in a celestial object's position depending on the location of the observer. In astronomy, this is usually caused by the movement of Earth, and is subdivided into diurnal parallax (caused by the Earth's daily rotation), annual parallax (caused by the Earth's movement as it orbits the sun) and secular parallax (caused by the movement of the solar system through space). The closer an object, the more it is affected by parallax—an effect observed in everyday life when roadside objects appear to move faster than distant ones viewed from a moving car.

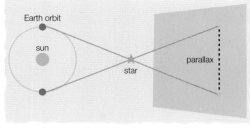

Parallax causes objects at different distances to appear to move relative to one another, even when they are not actually doing so.

proper motion

the angular movement of a star (or other object) on the celestial sphere. Proper motion is the portion of its actual movement that is perpendicular to the observer's line of sight.

radial velocity

the component of a celestial body's velocity directly along the observer's line of sight, usually calculated from the **shift** (red or blue) of the star's emitted light (or of the reflected light, for non-emitting bodies). Radial velocity is often measured relative to the sun, to avoid complications caused by the Earth's orbital motion, and is sometimes called the line-of-sight velocity.

space velocity

the overall velocity of a star (or other celestial body), equal to the sum of its **proper motion** and **radial velocity vectors**. The name is slightly misleading, as it is the velocity relative to the observer, rather than to space itself. (Celestial velocities can only ever be calculated relative to other celestial objects: there is no such thing as a fixed stationary point in space.)

Hubble constant

the rate at which distant objects move apart from one another due to the expansion of space, measured as a function of their current distance. The Hubble constant is a constant only in that it applies to all objects at the present time; it is believed to be decreasing with time. The best measurement of the current value to date is that the rate of expansion is between 67 and 75 km per second per megaparsec.

escape velocity

the minimum speed needed to move from the surface of an object to a position beyond its gravitational field. Theoretically, this position is an infinite distance away, although for practical purposes it is much closer. The Earth's escape velocity is just over 25,000 miles per hour (11,200 meters per second). It is not necessary for a vehicle to actually reach escape velocity if it is able to provide thrust after leaving the ground (as all current space vehicles do), as continued thrust allows an escaping vehicle to counteract the acceleration of gravity.

Lagrangian point

in a gravitational system with two large bodies, the points in space where the gravitational fields of those two bodies cancel out. There are five points: two are stable (an object nearby will tend to drift into the point) and the other three are unstable (so that an object not perfectly centered on the point will tend to drift away). The Trojan asteroids occupy the two stable points in the gravitational system of Jupiter and the sun. Space-stations on the Lagrangian points of the Earth–moon system are a popular feature of works of science fiction.

geostationary orbit

an orbit over the equator of a planet at a height such that the orbital period is exactly equal to the period of rotation of the planet, meaning that a satellite in such an orbit remains stationary relative to the planetary surface underneath. For Earth, the altitude of geostationary orbit is 22,500 miles (36,000 km). Geostationary orbits can only be achieved at other heights (or away from the Equator) by continuous thrust from the satellite to maintain its position relative to the ground.

❧ Roche limit

the minimum distance from the parent planet's center at which a satellite can orbit the planet safely. Within the Roche limit, the satellite will, depending on its relative mass, either spiral rapidly to the ground or be torn apart, forming rings (such as those of Saturn). If a planet and its moon are of similar density, the Roche limit is around 2.5 times the planet's radius.

Schwarzschild radius

the minimum radius of an object with specified mass at which the surface **escape velocity** is lower than the speed of light. Any star (or, in theory, other super-heavy object) that shrinks to such an extent that its radius is less than the Schwarzschild radius for its mass becomes a black hole. Once such an object has collapsed, the Schwarzschild radius becomes its **event horizon**.

event horizon

a boundary from beyond which no information can reach an external observer. Normally associated with a black hole, being

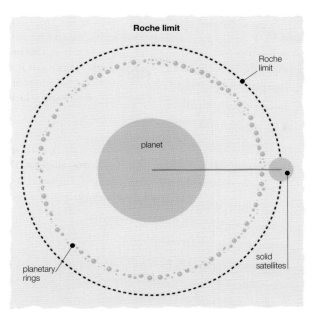

Roche limit

Roche limit

planet

planetary rings

solid satellites

The Roche limit varies with density; the denser the moon is relative to the planet, the closer it can come without being destroyed.

the distance from the center at which the gravitational field is sufficiently strong to prevent the escape of electromagnetic radiation (that is, the **Schwarzschild radius**). It is also possible to have non-gravitational event horizons, however—the **Hubble length**, for example, is the radius of a spherical event horizon around Earth.

quadrant

an instrument for measuring angles up to a maximum of 90°, and a predecessor to the sextant used for nautical navigation. The octant was a similar instrument, with a maximum measuring angle of 45°.

planisphere

a circular star chart showing all the celestial objects that can possibly be seen from a particular **latitude**, combined with an overlay marked with dates and times. The overlay is positioned on the chart according to the current date and time, with adjustment for the precise latitude and **longitude**, and allows the calculation of horizontal **celestial coordinates** (i.e., **azimuth** and **elevation**) for any visible object.

stereocomparator

a device for measuring the movement of stars and other celestial objects by comparing two photographs of the same part of the celestial sphere taken at different times (usually days or even months apart).

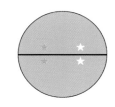

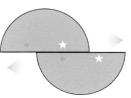

A simple model of how a heliometer works.

bifilar micrometer

a simple measuring instrument for finding the angular separation of two objects viewed through the same telescope using a pair of extremely fine parallel wires. Sometimes called a filar micrometer.

✥ heliometer

an instrument for measuring the **angular distance** between two objects, using a movable lens to generate a double image. The first image of one star is then aligned with the second image of the other by adjusting the lens, and the angular distance between them calculated from the adjustment required. It was originally designed for measuring the diameter of the sun, hence the name.

Langley

a unit of solar radiation, equal to an energy flux of one **calorie** per square centimeter (41.84 kilojoules per square meter). Often also used for this amount of energy per minute, which is approximately half the rate of solar radiation arriving on the Earth's surface.

jansky (Jy)

a unit of received power from a source of electromagnetic radiation, primarily used for the strength of signals received by radio telescopes. One jansky is equal to 10^{-26} watts per square meter per hertz of channel width. Thus, a signal with a channel width of 1 MHz would need 1,000 times as much absolute power as one with a channel width of 1 kHz to have the same rating in janskys.

shift (redshift, blueshift)

the change in the apparent wavelength of light (and other radiation) emitted by distant stars as seen from Earth. The two main causes of shift are relative movement (the **Doppler effect**) and the expansion of space. (These two causes are not the same—as a function of the expanding universe, the Earth's distance from other celestial objects is increasing because the space between is stretching, in addition to any relative movement of the two.) Redshift happens when an object is moving away from us; blueshift when it is coming closer. There is also gravitational shift, which occurs when light passes close enough to a massive object such as a star or black hole to be affected by its gravitational field.

spectral class

a measure of the temperature and chemical composition of a star, based on the intensity of light it emits at different **wavelengths**. As with solar flares (*see* p. 24), the class consists of a letter followed by a number between 1 and 9, sometimes with further information (such as a **luminosity class**) after the number. Most stars fall into one of the letter-classes (in order from hottest to coldest) O, B, A, F, G, K and M, but some unusual stars are assigned other letters, including W (very hot, with surface temperatures over 50,000 **kelvin**), C and S (similar temperatures to K and M, but with unusual chemistry). The numbers are a linear scale from hottest to coldest within each class, so an A4 star is hotter than an A6 one. The sun's spectral class is G2V.

luminosity class

a measure of a star's brightness, usually appended to its spectral class. Indicated with Roman numerals, on a scale theoretically from 1 to 7, though classes 6 and 7 are now rarely used. I indicates super-giants (the largest, brightest stars), while V is the much more numerous main sequence dwarf stars—such as the sun.

color index

a measure of the temperature of a star, calculated by comparing the intensity of its light in the blue and yellow color bands. The more blue light the star emits (relative to the amount of yellow light), the hotter it is.

magnitude

a measure of a star's brightness. The apparent magnitude of a star (or other celestial object) is its brightness as seen from Earth, defined on a geometric scale with lower numbers indicating brighter stars. The dimmest objects visible from Earth with the naked eye have a magnitude of six, and a reduction of each magnitude "point" on the scale makes an object 2.51 times brighter. The absolute magnitude of a star is the apparent magnitude it would have if seen from a distance of exactly 10 parsecs. However, planets, comets, or other similar bodies that are not light-emitting stars would be invisible at such a distance, so their absolute magnitude is calculated as the apparent magnitude the body would have if it were one astronomical unit (**AU**) from both Earth and the sun, and had a phase angle of zero. (Such a position is physically impossible, but is useful for calculating a theoretical absolute value.)

sunspot number (**Wolf sunspot number: R**)

a measure of the sun's surface activity as indicated by the presence of sunspots. Equal to the sum of the total number of visible sunspots plus 10 times the number of sunspot groups, multiplied by a factor (usually between 0 and 1) that depends on the position and type of the telescope used to make the observation. There are two "standard" sunspot numbers, calculated using the same formula but data from different sets of observatories. Of these, the National Oceanic and Atmospheric Administration's "Boulder" number is typically some 25 percent higher than the "International" number calculated by the sunspot index data center in Belgium.

solar flare intensity scale

a measure of the X-ray energy of a solar flare, and therefore the strength of the resulting disruption to radio and satellite communications. (The strongest flares can affect electronic equipment on the ground, including power transmission networks.) Solar activity is classified with a letter (A, B, C, D, M or X) and a number (normally from 1 to 9): A, B and C are the normal surface behavior of the sun and class D flares generally have no effect on Earth. Each letter class is 10 times more powerful than the previous one, with the numbers representing a linear scale within each class. Thus, an X3 flare is six times the power of an M5 flare, which is in turn five times the power of an M1 flare. The most powerful solar flare recorded to date occurred on November 4, 2003, with a rating of X28.

Solar flares are defined as a sudden, rapid and intense variation in brightness and occur when magnetic energy that has built up in the solar atmosphere is suddenly released. Large flares can emit up to 10^{32} ergs of energy per second, which is 10 million times greater than the energy released from a volcanic explosion. However, this amount is less than one-tenth of the total energy emitted by the sun every second.

24

Torino impact scale

a measure of the threat to human life posed by a comet, asteroid or other body approaching Earth. Events are rated on a scale of 0 to 10, subdivided into five classes. "Events having no likely consequence" (rating 0), and "events meriting careful monitoring" (rating 1) are low threat categories. "Events meriting concern" (ratings 2 to 4) are those where a collision is considered unlikely, and would only be likely to cause relatively local damage. "Threatening events" (ratings 5 to 7) are close encounters with significant possibilities of causing widespread (or even global) destruction, and ratings 8 to 10 are "certain collisions," ranging from relatively minor impacts causing purely localized devastation to global catastrophes capable of wiping out life on Earth.

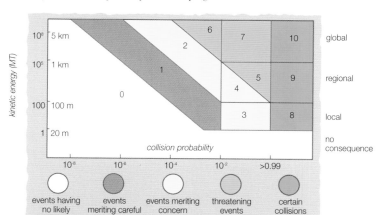

The Torino scale rates the danger posed by an object based on both the likelihood of collision and the amount of damage it would cause.

radiant

the point on the **celestial sphere** from which a meteor shower or similar phenomenon appears to originate when observed from the

Earth's surface. (It is actually a perspective effect of parallel lines, identical to the apparent narrowing of a straight road as it heads into the distance.)

shower length (shower unit; s)

the mean distance required for a charged particle (such as a cosmic ray) moving through a given medium to lose half its original energy. The shower length through air is around 250 yards (228 m), while through solid lead it is only 1.4 inches (35 mm). A related unit is the "cascade length," equal to the shower length divided by $\ln_2$ (about 1.44) where ln = "natural" **logarithm**.

epoch

in astronomical terms, a precise instant of time used as a point of reference for recorded data (e.g., the position of a star or planet in the sky). The combination of known starting time and location, and known movements, allows the calculation of the object's location at any other point in time. Where there is then a difference between the calculated and measured data, it may be used to correct assumptions or locate previously unknown objects whose gravity has affected the movement of the known object.

Eddington number

an estimate, produced by British astronomer Sir Arthur Eddington (1882–1944), of the total number of subatomic particles in the universe (about 1.5×10^{79} each of protons, neutrons and electrons). However, other than Eddington's own mathematical argument—based on measurements now believed to be incorrect—there is no evidence for the accuracy of this figure.

The Earth's orbit has an average radius of 1 AU. The average radius of Venus's orbit is about 0.7 AU, and for Mercury, about 0.38 AU.

✿ astronomical unit (AU)

a unit of distance, small by astronomical standards, equal to 149.6 Gm (93 million miles or 150 million km). Roughly equal to the mean distance from Earth to the sun, the AU is actually defined as the radius of a perfectly circular orbit with the same period as Earth.

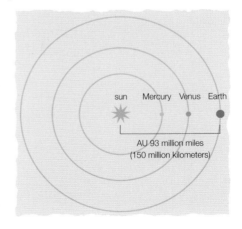

sun   Mercury Venus Earth

AU 93 million miles
(150 million kilometers)

light-year (ly)

the distance that light will travel through a vacuum in one year. Though the year can be calculated in different ways (astronomers generally use either **solar** or **sidereal years**, depending on the purpose), the difference is such a small proportion of a year as to be essentially unimportant. Different years produce light-years of between 63,240 and 63,280 **AU**. By extension, light minute, light second and so on are used to mean the distance that light travels in the specified time. It has been (jokingly) suggested that 1 light nanosecond (at 11.8 inches, very close to a foot) would be a more logical scientific unit of length than the meter, but no one wants to redesign the entire SI to fit a new base unit.

parsec (pc)

the standard astronomical unit of distance, defined as the radius of a circle in which 1 arc-second of circumference is 1 **AU** in length. One parsec is equal to 3.26 light-years, or just over 19.26 trillion miles. Proxima Centuri, the nearest star to Earth (other than the sun), is 1.29 parsecs (4.22 light-years) away.

Siriometer

a little-used unit of distance, with two possible interpretations. Originally intended as the distance from Earth to the star Sirius (just under 550,000 **AU**), but occasionally used to mean one million AU. It was also proposed as another name for the **parsec**.

Siriusweit

another ambiguous and little-used measure of distance. Occasionally used to mean five **parsecs** (about twice the actual distance to Sirius), on the grounds that it was once thought this might be the distance between Sirius and our sun. Other people have suggested that it is equivalent to the Siriometer, in either of its possible meanings, but the lack of a clear definition of the unit or any obvious need for it has left it almost totally disused.

galactocentric distance

for a named star, the calculated distance from that star to the center of the Milky Way galaxy. For the Earth's sun, this is approximately 8,500 parsecs, or 27,500 light-years—or 0.25 of a zettameter.

zettameter (Zm)

a very large unit of distance (10^{21} meters—around 106,000 light-years), used by astronomers for discussing the distances between galaxies. The nearest galaxy outside the Milky Way is thought to be the Canis Major dwarf galaxy (which orbits the Milky Way), the nearest edge of which is only around 25,000 light-years (0.24 Zm) from Earth. The Andromeda galaxy (the Milky Way's nearest major neighbor) is around 25 Zm distant, but approaching at a speed of some 310,700 miles per hour (500,000 kph). A collision between the two is expected in around two billion years' time.

Hubble length (LH)

the maximum distance that an object can be from Earth and still be visible. It is the number of light-years equal to the age of the universe in years. Objects 1 LH from Earth are moving away from us at the **speed of light**, and therefore at the limit of visibility.

Objects more than 1 LH apart logically have a relative velocity greater than the speed of light, a paradox which modern physics has yet to explain.

solar mass

a unit of **mass** (not weight), used to express the mass of stars being studied by astronomers. Defined as 1.989×10^{30} kg (2.2×10^{27} "short" **tons**), the approximate mass of Earth's sun.

ᔧ Jupiter mass

literally the **mass** of the planet Jupiter, used as a unit of mass by astronomers when discussing the planets in orbit around other stars. Approximately 1.9×10^{24} metric tonnes, or just under a thousandth of the mass of the sun.

Earth mass

the **mass** of our homeworld, just under 6×10^{21} metric tonnes (6.6×10^{21} federal or Customary tons—one Jupiter is around 315 Earths). Used as a comparative unit by astronomers looking for similar planets in other star systems. Although a number of Jupiter-size planets are now known, Earth-size planets are much harder to locate and none has been definitely confirmed orbiting other stars, though several possible candidates have been found.

Jupiter is by far the heaviest planet in the solar system, but still (at 1.9×10^{24} tonnes) has less than 0.1 percent of the mass of the sun.

ᔧ moon, phases of

the change in shape of the visible part of the moon as its phase angle changes. The visible proportion of the moon's surface increases as the phase angle decreases, with full moon when the phase angle is very close to zero and a lunar eclipse when it is exactly zero. Planets, viewed through a telescope, have phases in exactly the same way.

synodic period

the time taken for an object to return to the same position in the sky as seen from Earth (relative to the sun). The synodic month, for example, is the exact time between two consecutive full moons.

The moon is referred to as waxing as it grows from new to full, and then waning as the visible area decreases again. A gibbous moon is one that is more than half-full—whether waxing or waning.

solar cycle

the length of time between two consecutive periods of maximum solar activity (when sunspots and **solar flares** are most common). The average period is approximately 11 years but cycles as short as nine and as long as 14 years have been known.

saros

a unit of time used for the prediction of eclipses, equal to exactly 223 synodic months, or—using the **Gregorian calendar**—18 years, 11 or 10 days (depending on the number of leap years) and 7.4 hours. After this length of time, Earth, sun and moon have returned to the same positions relative to one another. The extra hours, however, mean that each eclipse will be visible from a different part of the Earth's surface, returning to approximately its original position every three saros.

Metonic period

an ancient approximation to the **saros**, used by the Hebrew and Babylonian calendars among others. The Metonic cycle repeats every 235 synodic months, equal to 19 **tropical (solar) years** and approximately two hours. It is used to calculate the date of Easter, and to determine the number of months in a particular Jewish year.

Dionysian period

532 years, in theory the length of the repeating pattern of the date of Easter. This is the lowest common multiple of the **Metonic period** and the 28-year repeating pattern of days of the week in the **Julian calendar**. Its practical value is limited, however, as the Metonic period is not precisely 19 years, and the **Gregorian calendar**'s lack of leap years on most centenary years means that the 28-year cycle is not always fixed.

platonic year (great year)

the time taken for the Earth to make one complete precessional rotation (i.e., for the **precession** of the equinoxes to return the equinoctial points to their original positions on the ecliptic). It equals approximately 25,800 Earth years.

❧ galactic year (cosmic year)

the time taken for the sun to make one complete orbit of the Milky Way galaxy. Generally reckoned to be approximately 225 million Earth years.

age of solar system

estimated to be around 4,600 million years. This is the time since significant solid parts of the sun and planets formed out of the pre-existing gas cloud. The estimate is based on radio-isotope dating of meteorites, ancient Earth rocks and rocks from the moon, all of which give virtually identical results.

ଈ age of universe

the time calculated (from measurements of distant stars) to have elapsed since the "big bang." The current best guess—based on data from the Hubble space telescop— is just under 14 billion years, but there are various estimates, ranging from around 13 billion to 20 billion years.

zodiac

a set of 12 constellations, all approximately on the ecliptic, used by astrologers to indicate the relative positions of the various planets (along with the moon and sun). Modern astrology tends to work by dividing the ecliptic into 12 equal parts, while astrologers in ancient times used the actual boundaries of the constellations (which are not all of equal width). The 12 signs of the western zodiac are Aries, Taurus, Gemini, Cancer, Leo, Virgo, Libra, Scorpio, Sagittarius, Capricorn, Aquarius and Pisces, but other astrological systems have different zodiacal signs with often entirely contradictory meanings attached to them.

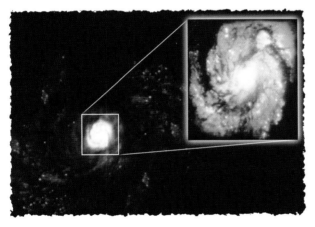

Galaxies are tens or hundreds of thousands of light years across, but look like stars to the naked eye. This one (NGC1275), in the constellation of Perseus, is a fairly near neighbor—only around 300,000 light years away.

29

age (of Pisces, Aquarius, etc.)

an astrological period of time. Depending on the definitions used, this is either 2,100 years (the time the **precession** of the equinoxes takes to move 30° round the ecliptic), or the period for which the vernal equinox actually occurs within a particular zodiacal sign. This disagreement of meaning leads to argument among astrologers as to whether we are currently in the age of Pisces (in which the vernal equinox will remain until around C.E. 2600) or the age of Aquarius.

ascendant

the astrological sign of the zodiac that was rising over the horizon at the time and place of a person's birth.

midheaven

the astrological sign of the zodiac that was closest to **zenith** at the time and place of a person's birth. Since the signs of the zodiac are not of precisely equal widths, it is usually but not always the sign three ahead of the ascendant for the same time and place.

house

a set of 12 divisions on a person's horoscope. Depending on the astrologer, these may be of even or varying widths, assigned to a zodiacal sign or relative to the precise moment of birth.

opposition

a term used to describe the position of two objects that have ecliptic longitudes 180° apart. However, they are rarely actually opposite one another in the sky, which would require that they also have equal and opposite ecliptic latitudes. They may in fact be very close to one another, although this is not common in astrology, since the objects dealt with usually remain within about 8° of the ecliptic.

trine

like **opposition**, a unit of relative position but with a difference of 120° between the two objects' ecliptic longitudes.

quadrate

as **trine** and **opposition**, a unit of relative position but with a difference of 90°. (Sometimes also called "square.")

conjunction

the opposite of **opposition**. Two planets (or a planet and the sun) are said by astrologers to be in conjunction when they appear close to one another in the sky.

❧ Chinese astrological cycle

a more complex arrangement than its western counterpart. Chinese astrology has 12 animal signs (though these are not strictly a zodiac) and five "elements" (similar to the four elements of western alchemy). These combine to give a cycle of 60 years.

In the Chinese astrological cycle animals have their own "inherent" elements, too: Tiger, Rabbit (Wood); Snake, Horse (Fire); Ox, Dragon, Sheep, Dog (Earth); Monkey and Rooster (Metal); and Rat and Pig (Water). The Rabbit is sometimes the Cat, and the Sheep sometimes the Goat. The cycle progresses across each row in turn; thus the year of the wooden Horse follows that of the water Snake.

wood		fire		earth		metal		water	
yang	*yin*	*yang*	*yin*	*yang*	*yin*	*yang*	*yin*	*yang*	*yin*
rat	ox	tiger	rabbit	dragon	snake	horse	sheep	monkey	rooster
dog	pig	rat	ox	tiger	rabbit	dragon	snake	horse	sheep
monkey	rooster	dog	pig	rat	ox	tiger	rabbit	dragon	snake
horse	sheep	monkey	rooster	dog	pig	rat	ox	tiger	rabbit
dragon	snake	horse	sheep	monkey	rooster	dog	pig	rat	ox
tiger	rabbit	dragon	snake	horse	sheep	monkey	rooster	dog	pig

distance

metric mile

a unit of length mostly used in athletics. A metric mile is different from an imperial, or statute, mile. Principally used as a race distance, it is generally accepted to be equivalent to 1,500 meters, or approximately 0.932057 imperial miles. Occasionally it is applied to a distance of 1,600 meters (which is in fact much closer to an imperial mile); this usage is common in U.S. high-school races.

ᘒ kilometer (km)

a unit of length equal to 1,000 meters, or 0.621371 imperial miles. The kilometer is the largest multiple of the meter in common usage.

meter (m)

the basic unit of length in the metric system. A meter is made up of 100 centimeters, or approximately 39.37 inches. The word derives from the Greek μετρον (*metron*), and its definition has been a subject of much scientific discussion over the past 200 years. According to the Conférence Générale des Poids et Mesures (a body of delegates all of whom have signed the Meter Convention and which is concerned with regulating and improving the metric system), it is currently accurately defined as the distance light travels in a vacuum in $1/299,792,458$ seconds. This may change in the future as scientists make more accurate estimates as to the speed of light.

prefix	abbrev	factor
kilo-	k	10^3
hecto-	h	10^2
deka-/deca-	da	10
—	—	1
deci-	d	10^{-1}
centi-	c	10^{-2}
milli-	m	10^{-3}
micro-	μ	10^{-6}
nano-	n	10^{-9}

Some common metric prefixes. These are placed before units of metric measurement to indicate that the unit has increased by a certain factor.

ᘒ centimeter (cm)

a unit of length equal to $1/100$ of a meter or approximately 0.39 inches.

ᘒ millimeter (mm)

a unit of length equal to $1/1000$ of a meter, or approximately 0.039 inches.

ᘒ micrometer (μm)

a unit of length equal to one-millionth of a meter, or approximately 0.00004 inches. This unit of length was first defined as a micron by the Conférence Générale des Poids et Mesures in 1879. In 1967 the C.G.P.M. declared the term micron to be obsolete, and replaced it with the micrometer. However, the word micron still exists as the more common term for this unit of length, especially in the scientific community where it is used to describe the length of particles or when referring to the wavelengths of light. A micrometer is also the term for an instrument used to measure small distances or thicknesses.

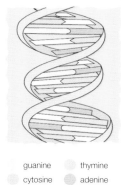

guanine   thymine
cytosine   adenine

The DNA helix is so small that it is measured in nanometers. One nanometer is 10⁻⁹ meters.

micron *see* micrometer.

﹐ nanometer (nm)

a unit of length equal to one-thousand-millionth of a meter, or approximately 0.00000004 inches. This unit was initially defined as a millimicron (mμ), but the term nanometer came into use when the micron was replaced by the micrometer. The nanometer is used, among other things, to measure wavelengths, such as those of visible light, gamma rays and ultraviolet radiation. Einstein calculated the length of a sucrose molecule to be approximately 1 nanometer.

angstrom (Å)

a unit of length equivalent to 10^{-10} meters, or 0.1 nanometers. The unit was first used by the Swedish physicist Anders Jonas Ångström (1814–74) to enable him to describe the solar spectrum. Although it is still occasionally used to describe the radii of atoms, which are between 0.25 and 3Å, it is more usual now to define such lengths in terms of nanometers. As such, the angstrom is largely obsolete; indeed its use is discouraged by the Conférence Générale des Poids et Mesures, and it is not an accepted SI unit.

﹐ femtometer (fm)

a unit of length equivalent to 10^{-15} meters. This unit of length was originally named the fermi, after the Italian–American physicist Enrico Fermi (1901–54), and although the femtometer is the accepted SI unit, physicists still refer to it as a fermi. The femtometer is used to measure the size of the nuclei of atoms, as well as that of sub-nucleic particles—protons and neutrons are approximately 2.5 femtometers in diameter.

fermi (fm) *see* femtometer.

Planck length (lp)

a unit of natural length used in quantum mechanics and astronomy. First defined by the German theoretical physicist Max Planck (1858–1947), it is the smallest unit of length defined by modern theoretical physics. It is not possible accurately to state the size of a Planck length in terms of standard measurements, but it is thought that it may be as small as a millimeter divided by a hundred thousand billion billion billion. A particle accelerator that could probe distances that small would be larger than our solar system; to probe distances smaller than a Planck length would require particles that can only be found inside black holes.

inch (in)

an imperial unit of length equivalent to 2.54 centimeters, or $^{1}/_{36}$ yard. The inch is a very ancient measurement. It is thought that it was first defined as the distance between the tip of the thumb and the knuckle of the thumb. Indeed, in certain languages, the words for inch and thumb are very similar. The U.S. survey inch is fractionally larger than the standard imperial inch, the difference between the two only becoming significant when they are used to describe distances of many thousands of kilometers.

foot (ft)

an imperial unit of length equivalent to 12 inches, or 30.48 centimeters. It is popularly thought that the foot was originally defined by the length of a human foot. Human feet are generally smaller than 12 inches in length, however. In the U.S. it is more properly referred to as the international foot, to distinguish it from the U.S. survey foot (*see* **inch**).

The diameter of nuclei of atoms is generally measured in femtometers.

yard (yd)

an imperial unit of length equivalent to 3 feet or 0.9144 meters. The yard has been standardized in a number of ways throughout history. Legend has it that Henry I of England defined it as the distance between the tip of his thumb and the tip of his nose with his arm outstretched; another theory states that it derives from the girth of a person's waist. "Yardsticks" of varying lengths have been made throughout history, but it is now accurately calibrated against the **meter**.

rod

an imperial unit of length, traditionally employed to measure land. Equivalent to 16.5 feet or 5.03 meters, the rod is now only really used as a unit in North America, and it is also sometimes referred to as a perch (as distinct from a *perche*) or a pole. Historically the rod was a much less specific measurement. Records from eighth-century Britain define an acre of land as being 40 rods long by 4 rods wide, but the length of an acre was also defined as the distance a team of oxen could plow without needing a rest, and its width as the number of lengths a team of oxen could plow in a day.

perche

a unit of length and area, or both, having different definitions in different countries. The *perche* is quite distinct from the U.S. perch (see **rod**), and is defined in specific countries as follows: Canada, 231.822 inches, the Seychelles, approximately 6.497 meters; Switzerland, 3 meters; Belgium, 6.5 meters. In pre-metric France the *perche* was an important land measurement which had a number of different definitions according to the region.

chain

a non-metric unit of length used by surveyors. More properly known as Gunter's chain, and most commonly used in U.S. public land surveys, it is equivalent to 22 yards or 20.1168 meters, and is divided into 100 links. In Scotland and Ireland, Gunter's chain is of different, much shorter, lengths: 8.928 inches in Scotland, 10.08 inches in Ireland. There are two other types of chain. Ramden's and Rathborn's, but these are much less common. In Cyprus, a chain is 8 inches long.

furlong

an imperial and U.S. customary unit of length equivalent to 660 feet or 201.168 meters. The word furlong derives from the Old English words *furh* (furrow) and *lang* (long), and it historically referred to the length of a furrow in a common field of 10 acres. There are 8 furlongs in a mile, its modern usage deriving from the Roman measurement of a stadium, which was $1/8$ of a Roman mile. The furlong's use is now largely restricted to British horse racing.

mile

an imperial unit of length also known as the international or statute mile. It is equivalent to 1,760 yards or approximately 1,609 meters. The nautical mile (also known as the "Admiralty mile"), however, is equivalent to 1,853 meters, and is used to navigate at sea and in the air. The statute mile was defined by the English Queen Elizabeth I as 8 furlongs in 1593, but the word itself derives from the Latin *mille passus* or **Roman mile**).

league

a unit of distance by land. The league has historically been used around the world to describe an approximate distance, but it has now fallen into disuse practically everywhere. Originally defined as being the distance a man or a horse can walk in an hour, it became an accepted length of roughly 3 miles around the 16th century, though with several regional variations.

seven-league boots

a fictional garment allowing the wearer to cover great distances in a single stride. Seven-league boots appear throughout the literature of European folklore.

cable

a nautical unit of distance, having a number of varying definitions. It is commonly accepted as being $1/10$ of a nautical mile, or 185.3 meters. Historically, a cable is 100 fathoms, or 182.88 meters. In the U.S. Navy, a cable is 120 fathoms, or 219.456 meters, whereas in the British Royal Navy, a cable is 608 feet, or 185.3184 meters. The cable is sometimes called the cable length.

fathom

a nautical unit of distance. A fathom is equivalent to 6 feet or 1.8288 meters, although it was originally deemed to be the length of a man's arm span. Historically used for distances across land and water, it was latterly reserved solely for measuring the depth of water. The fathom fell out of official nautical usage in 1999.

sounding

the process of measuring the depth of water using a sounding line or an echo system. A sounding line is a weighted line with distances marked at regular intervals. It is hung from the side of a boat to determine the depth of the water underneath. An echo system achieves the same result by bouncing a sound wave from the boat off the bottom of the body of water and measuring the time it takes for the sound wave to return to the transmitter.

span

a unit of length equivalent to 9 inches or 22.86 centimeters. Historically it was taken to be the distance between the tips of the thumb and little finger when outstretched.

hand (hh)

a unit of length equivalent to 4 inches or 10.16 centimeters. Originally considered to be the breadth of a hand, it is now only used to measure the height of horses. The abbreviation hh stands for "hands high."

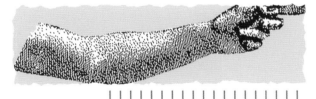

The cubit is one of the most ancient of all units of distance, but the word has been used to represent different distances throughout history and around the world.

❧ cubit

a very ancient unit of length used by several civilizations. The word derives from the Latin *cubitum*, meaning elbow, and was approximately equal to the length of a man's forearm. The Roman cubit was a length of about 44.35 centimeters, but the cubit also existed as a unit of measurement in the ancient kingdoms of Babylonia and Egypt. The Babylonian—or Sumerian—cubit was 51.72 centimeters, and this is the earliest known standard measurement of length. The Egyptian cubit is thought to have been nearer 52.4 centimeters. The cubit was used as a unit of measurement equivalent to 18 inches throughout the British Commonwealth, but it has now fallen out of use.

digit

a Roman unit of length. Traditionally thought to be the width of a man's finger, the digit is believed to have been approximately 18.48 millimeters.

uncia

a Roman unit of length and weight. In length. it was $1/12$ of a *pes*. or approximately 24.6 millimeters. As a unit of weight. it was $1/12$ of a *libra*. or 27.3 grams. It is from the Latin *uncia* that the English words "inch" and "ounce" derive. and *unciae* were also coins of this weight.

palm

a unit of length equivalent to 3 inches or 7.62 centimeters. although it had previously been a less specific measurement referring to the breadth of the hand at the base of the four fingers. not including the thumb. It derives from an earlier Roman unit of length. the *quadrans*. which was equivalent to three *unciae* or approximately 7.38 centimeters.

pes

a Roman unit of length equivalent to 12 *unciae* or approximately 295.7 millimeters. The Latin word *pes* translates into English as "foot." and indeed the *pes* is equal to 11.6 inches—not far off an imperial foot.

pace

a unit of length derived from the Roman *passus*. which was equal to 5 *pes*. or 147.85 centimeters. The English pace became standardized at 5 feet. or 152.4 centimeters. although the American pace became independently standardized at just half this length.

stadium

an ancient Roman or Greek unit of length. The English word stadium derives via Latin from the Greek στάδιον (*stadion*). and this unit of measurement corresponded to about 606 feet (185 m). The oval arenas in which the Greeks and Romans held their sports were generally built to this length. and so the unit of measurement gradually became the word for the arenas themselves.

Roman mile

an ancient Roman unit of length from which the modern statute mile is derived. The Latin for the Roman mile was *mille passus*. meaning "a thousand paces." but in fact the *passus* itself was a specific measurement equal to 5 *pes*. It is accepted that the Roman mile corresponded to approximately 1.485 meters.

marathon

a long-distance running race. The marathon commemorates the myth that after the Greeks overcame the Persians in battle in 490 B.C.E.. a messenger ran from Marathon to Athens—a distance of about 22 miles—with the news. (In fact. the Greek historian Herodotus gives an even earlier version of the story. in which the messenger Pheidippides ran not 22 but 150 miles from Athens to

Sparta to seek help before the battle.) The earliest marathon race was held at the first modern Olympic Games in 1896, but the length of the race was not standardized by the International Olympic Committee until 1924, when it was decided that it should be a length of 26 miles 385 yards, or 42.195 kilometers.

standard gauge

a standardized distance between the two rails of a railway track. Approximately 60 percent of the world's railway tracks are built to this gauge, which is 4 feet 8^1/2 inches or 1.435 meters. This gauge was standardized in Britain by the Gauge Act of 1846—previously railway tracks had been set at the slightly narrower width of 4 feet 8 inches—and the rest of the world soon followed suit. Broad gauge, which is 7 feet ^{1}/4 inch or 2,140 mm wide, is used by high-speed trains in continental Europe.

❧ caliber

the internal diameter of a gun barrel, or the diameter of a bullet or other piece of ammunition. Gun calibers can be described in inches or millimeters. Inches tend to predominate in North America and the U.K. So, a .44 caliber gun is a gun whose barrel has an internal diameter of .44 inches. Similarly, a 9 mm revolver has a barrel whose internal diameter is 9 millimeters.

bore

the internal diameter of a cylinder. The bore of a cylinder is measured in inches or millimeters; when referring to shotguns (and some other weapons), however, the bore is defined quite differently: it is the number of solid rounds for the weapon that can be made from 1 lb of lead. Hence a 4-bore shotgun is a more formidable weapon than a 12-bore.

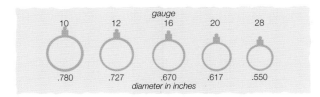

The diameter of a gun barrel is measured in terms of its caliber. Here, five calibers are illustrated: .78, .727, .67, .617 and .550.

calipers

an instrument for measuring dimensions. Resembling a pair of compasses, a set of calipers consists of two hinged legs with sharp points. These points are either directed outwards (to measure internal distances) or inwards (to measure external distances).

❧ vernier scale

a moveable scale on the main scale of a measuring device. Devised by the French mathematician Pierre Vernier (1580–1637), a vernier scale enables the user to obtain fractional parts of the sub-divisions on the main scale of a measuring device. Vernier scales can be found on. for example. barometers. sextants. calipers and micrometers.

odometer

an instrument for measuring the distance traveled by a wheeled object. The most common type of odometer is the dial on a car dashboard which states the distance the car has driven. although these are being replaced increasingly by digital odometers. However. odometers—or at least the idea of the odometer—have been around for many thousands of years. Reference is made to an odometer by Vitruvius around 25 B.C.E.. and it is thought that Archimedes may have invented the first one. Leonardo da Vinci attempted to build an odometer as described by Hero of Alexandria. but failed.

pedometer

an instrument for estimating distance traveled on foot. The user estimates their average stride length. and the pedometer measures the number of strides by means of a component that moves with the jolt of the user's step. The distance traveled is equal to the average stride length multiplied by the number of strides.

line

a small linear measure. now rarely used. equivalent to $1/12$ of an inch (2.1167 millimeters). When used to measure the thickness of buttons or watch movements. a line is an even smaller measure equivalent to $1/40$ of an inch (0.635 millimeters).

The vernier scale allows small measurements to be subdivided into even smaller measurements.

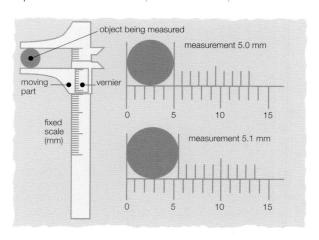

geology

❧ soon

changes in geology happen very slowly. Major changes take millions of years, so the notion of "soon" to a geologist is likely to be anything less than a million years. Similarly, "recently" is a comparative notion. One way to understand it is to imagine that Earth is a year old, and that the first multicellular life forms appeared in May or June, plant life in August, dinosaurs in late December, and humans between 11:30 p.m. and midnight on December 31.

era	period	epoch	m.y.a.	notes
Hadean			4,500	Earth forms as a solid planet. No evidence of life.
Archaean			3,900	Solid crust forms. Earliest single-celled life.
Proterozoic			2,500	Mountain ranges begin to form. Earliest multicellular life. Movement of tectonic plates slows to about the current rate.
Paleozoic	Cambrian		570	Metazoans (sponges and corals) and trilobites appear. Supercontinent Gondwana begins to break up.
	Ordovician		505	Earliest fish, but most life still invertebrate. No life out of water.
	Silurian		435	First sharks appear, along with the first plants on land.
	Devonian		405	Ammonites, amphibians and first air-breathing arthropods.
	Carboniferous		365	First flying insects appear, plant-life well established on land. Late in this period the first reptiles appear.
	Permian		290	Trilobites extinct. Supercontinent Pangaea formed.
Mesozoic	Triassic		245	First dinosaurs and earliest mammals appear on land.
	Jurassic		210	Birds evolve. Pangaea breaks up; the Atlantic forms.
	Cretaceous		145	First flowering plants. Dinosaurs and ammonites wiped out. Continents have similar forms to today, but different positions.
Cenozoic	Tertiary	Paleocene	65	Inland seas dry up. Ungulates, rodents and primates evolve.
		Eocene	57	Alpine-Himalayan and Rocky mountains begin to form.
		Oligocene	35	Grasslands expand at expense of forests. First apes appear.
		Miocene	25	Higher primates evolve. Climate cools, Antarctica freezes.
		Pliocene	5	First appearance of a species classified as "homo."
	Quaternary	Pleistocene	1.75	First appearance of *Homo sapiens*.
		Holocene	0.015	Earliest known civilization begins.

age of Earth

the oldest known Earth rocks date from around 3.8 to 3.9 billion years ago, but contain minerals that are 4.1–4.2 billion years old. These ages are established by **radio-isotope dating** and by studying meteorites believed to have formed at the same time as the planets.

radio-isotope dating

radio-isotope dating relies on finding out how much an isotope has decayed over the years into its "daughter" isotope. For example, zircon crystals, which form in cooling magma, capture radioactive uranium-235, but not lead. Because there was no lead initially, and uranium-235 decays to lead-207, the amount of lead there now is

The geochronological table. Pre-Cambrian time consists of three eras: the Hadean, the Archaean and the Proterozoic. Earth is believed to have been formed in the Hadean era 4,500 million years ago (m.y.a.) as a solid planet, but with no evidence of life.

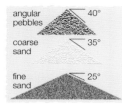

Different loose materials have different angles of repose. If sand, for example, is piled up into a heap, grains will constantly slide down its surface until equilibrium is reached, at which point the angle of the slope is the angle of repose.

The volcanic explosivity index (V.E.I.) measures the explosivity of an eruption, the volume of ash (tephra) and the height of the resulting ash cloud column. During human history no volcanic eruption has reached 8 on the V.E.I. scale, though this does not mean one of such magnitude has not occurred during the geologic past.

an indication of the rock's age. The half-life of uranium-235 is 704 million years, so half the uranium-235 atoms originally present in a sample of zircon will be lead-207 after 704 million years. Other isotopes used for dating rock include rubidium-87 (half-life: 48,800 million years; daughter: strontium-87) and potassium-40 (half-life: 1,280 million years; daughter: argon-40).

carbon dating

radio-isotope dating using an isotope of carbon, carbon-14, whose "daughter" isotope is nitrogen-14. Carbon dating is useful because carbon-14's half-life is relatively short (in geological terms!)—just 5,730 years—so is good for dating items between 500 and 50,000 years old. As with all radio-isotope dating, care has to be taken to ensure that a sample has not been contaminated, for example by carbon-dioxide emissions from volcanic eruptions.

clinometer

an instrument for measuring the angle of inclination in rock **strata**. Movements of the Earth's crust often cause layers of rock that were originally laid down horizontally to slope at a significant angle.

?▲ angle of repose

the greatest angle of slope that loose material can sustain.

geothermal energy

heat energy from below the Earth's surface, particularly in areas of volcanic activity, seen as a form of renewable energy, as it is always there and in theory may be tapped constantly. The hottest parts of the Earth's surface are close to active volcanoes, but areas of little or no volcanic activity can produce significant amounts of energy.

?▲ volcanic explosivity index

a scale used to gauge the intensity of volcanic eruptions. It also includes a description of the eruption and a name for it. The number of eruptions in history is also recorded.

volcanic explosivity index	0	1	2	3	4	5	6	7	8
general description	non explosive	small	moderate	moderately large	large	very large ————————→			
volume of tephra (m³)		10^4	10^6	10^7	10^8	10^9	10^{10}	10^{11}	10^{12}
cloud column height (km)*	<0.1	0.1-1	1-5	3-15	10-25	25 ————————→			
qualitative description	gentle, effusive	←—— explosive ——×——		cataclysmic, paroxysmal, colossal —→ severe, violent, terrific ————————←					
classification		←—— Strombolian ——×—				—— Plinian ————————→			
	Hawaiian	←———— Vulcanian ————×——			ultra-Plinian ————→				
total historic eruptions	487	623	3,176	733	119	19	5	2	0
1975–85 eruptions	70	124	125	49	7	1	0	0	

* Note: For VEI's 0-2 data are in km above crater, for VEI's 3-8, data are in km above sea level.

volcanic activity

Volcanoes are usually described as extinct, dormant or active. Extinct means that all volcanic activity has now ceased; dormant means there is definitely some volcanic activity, but that the volcano has not erupted for a long time, and shows no sign of imminent eruption; active means that it is likely to erupt at any time. A notable example of an extinct volcano is Mount Fuji in Japan, and a particularly active volcano is Mount Etna in Italy. Mount Lassen in California, which last erupted in 1915, still "murmurs" occasionally, and so is not extinct but dormant.

fault

a part of the Earth's crust where there is a gap or crack caused by a displacement of one side against the other. The line along the crack is referred to as the "fault line" and the plane in which movement takes place is the "fault plane."

tectonic drift

the Earth's crust is made of a series of sections, called tectonic plates, which constantly move against one another, causing major natural features of the landscape, particularly mountain ranges. They typically move at up to about 2 inches (5 cm) a year. There are seven major tectonic plates, and a number of smaller ones.

strata

the individual layers of sedimentary rock, laid down over time. The layers of different types of rock are often visible, e.g., on cliff faces.

✍ Richter scale

a measure of earthquake magnitude, named after American seismologist Charles Richter (1900–85). It is a logarithmic scale, beginning at 0, with each subsequent whole number representing ten times the magnitude of the previous one, and about 32 times the amount of energy released. It differs from the **Mercalli scale** in that it uses data recorded on seismographs, not observed effects of an earthquake on buildings and other structures. The earthquake in the Indian Ocean in December 2004 measured 9.3 on the Richter scale, and as such was one of the largest ever recorded.

✍ Mercalli scale

a scale of the effects ("intensity") of earthquakes. Named after Italian seismologist Giuseppe Mercalli (1850–1914), who built on the work of physicists de Rossi and Forel to devise a scale indicating the observed effects of earthquakes on the ground—from 1 (effects only detected by seismographs) to 12 (total destruction). Further modified by American seismologists Harry Wood and Frank Neumann, the scale is now usually called the modified Mercalli scale.

Mercalli scale	
1	Not felt by people.
2	May be felt by people at top of buildings. Hanging objects begin to swing.
3	Vibration like passing small truck. Hanging objects set in motion. May not be recognized as earthquake.
4	Vibration like passing of heavy truck. Standing cars may rock, dishes rattle.
5	Felt outdoors. Liquids slop out of cups. Small objects may topple over.
6	Felt by all. People may be frightened. Pictures fall off walls. Glass may crack.
7	Difficult to stand. Weak chimneys break. Fall of plaster, tiles, cornices. Waves on ponds.
8	Steering of cars affected. Fall of some masonry walls. Severe structural damage to buildings. Changes in flow or temperature of springs.
9	Large-scale damage to buildings, dams, embankments. People panic, animals run.
10	Most buildings destroyed. Landslides, water thrown from rivers.
11	Destruction of roads, rails and underground services. Large cracks in ground.
12	Damage total. Large rock masses and water courses displaced. Lines of sight and level distorted. Objects thrown into the air.

✿ moment magnitude scale

a successor scale to the **Richter scale** for measuring the magnitude of earthquakes. The scale was developed by Hiroo Kanamori, a Japanese seismologist who recognized that the Richter scale had a problem of "saturation" at high values, i.e., the higher up the scale, the less distinction between the magnitudes of earthquakes. Hence, there could be significant differences in energy released by earthquakes ostensibly measuring about the same on the Richter scale. The moment magnitude scale takes account of low-frequency seismic waves, which can often cause the most damage to large buildings.

Comparison between the Richter and Moment Magnitude scales		
earthquake	*Richter scale*	*Moment Magnitude scale*
New Madrid, MO, 1812	8.7	8.1
San Francisco, CA, 1906	8.3	7.7
Prince William, AK, 1964	8.4	9.2
Northridge, CA, 1994	6.4	6.7

sedimentation rate

the rate at which particles suspended in water or air settle. Natural sedimentation occurs at different rates with different materials and conditions, so measurements of sedimentation for particular materials can give an indication of how conditions have changed over time.

Zhubov scale

a scale for measuring ice coverage. Devised by a Soviet naval officer, N.N. Zhubov, the scale uses a unit called a ball. Clear water is 0 balls, 10 percent ice coverage is 1 ball, 20 percent is 2 balls, etc.

✿ latitude

an imaginary line circling the globe parallel to the Equator. The Equator has a latitude of 0 degrees, and there are 90 degrees between the Equator and each of the Poles. The Tropic of Cancer lies at 23.5° north, and the Tropic of Capricorn at 23.5° south. Each degree represents about 69 miles (111 km) in distance at ground level.

✿ longitude

an imaginary line, also known as a meridian, along the surface of the Earth between the North and South Poles. Zero degrees longitude is the Greenwich or prime meridian, passing through Greenwich in London, England. The line of longitude farthest from the Greenwich meridian is 180° east or west, which is also essentially the location

The imaginary lines crisscrossing the Earth's surface provide a grid reference system to identify the location of any place.

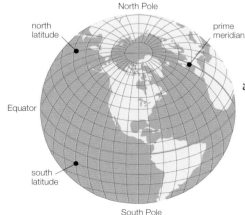

North Pole

north latitude

prime meridian

Equator

south latitude

South Pole

of the International Date Line. With **latitude**, longitude forms part of a grid reference system that can identify the position of any place on Earth.

North and South Poles

the Poles are the points at maximum distance from the Equator in the Arctic and Antarctic respectively, and the points of convergence of lines of **longitude**. Earth has a magnetic field whose poles roughly coincide with the North and South Poles, so adjustment must be made to compass readings to get true bearings—the farther north you go the more important this is. Magnetic North is about 625 miles (1,000 km) from True North, and is constantly moving at 6–25 miles (10–40 km) a year. The polarity of the Earth's magnetic field has completely reversed many times in the past.

field capacity

the amount of water contained in soil after normal drainage has finished. In other words, the water content after natural drainage due to gravity and without more water being added. The remaining water in the soil is known as the capillary water.

strike and dip

two words to describe the angle of inclination to the horizontal of layers of rock or a fault. Strike indicates the direction of a line that is formed where the inclined plane and the horizontal intersect. It is the angle between this line and True North. Dip is the tangent to the surface of Earth at the point where the dipping surface cuts it.

gravimeter

a gravimeter is an instrument that measures Earth's gravitational field at different points on the surface, allowing comparison to be made between the different points. It is particularly useful in prospecting for oil and minerals. One simple form of gravimeter is a weight suspended from a sensitive spring: the greater the gravitational pull, the more the spring will be stretched.

inclinometer

An inclinometer measures dip (*see* **strike and dip**). A simple form of inclinometer is a weight suspended from a string attached to the center of a protractor.

magnetic declination (magnetic deviation)

quite simply, the difference between the direction of True North and Magnetic North (*see* **North and South Poles**). Because of the constant change in location of Magnetic North, the magnetic declination is also constantly changing.

isodynamic lines

imaginary lines on the Earth's surface connecting points of equal magnetic intensity. Also sometimes referred to as isomagnetic lines.

land area

hectare

decare

are

deciare

1 km

1 km

The hectare is 1 square kilometer. One hectare consists of 10 decares, 100 ares or 1,000 deciares.

❧ hectare (ha)

a metric unit of land area equal to one square kilometer or approximately 2.471 acres. Derived via French from the word *are* and the Greek *hekaton*, it is now in more common usage in the U.S. than elsewhere in the world. It is not an official SI unit, but its use is accepted under that system, though discouraged.

are (a)

a seldom-used metric unit of land area. The are is equivalent to 100 square meters, $1/100$ of a hectare or 1076.3910 square feet. The word derives, via French, from the Latin *area*, meaning a level, open space.

decare

a metric unit of land area equivalent to 10 ares. Ten decares, therefore, are equal to one hectare, but the measurement is little used.

deciare

a metric unit of area equivalent to 0.1 of an are, or 10 square meters. The prefix "deci" is often used to indicate one-tenth of a particular measurement.

square meter (m^2)

the fundamental metric unit of area. A square meter is the area enclosed by a square whose sides are each 1 meter in length. Just as the meter is the basic SI unit of length, so the square meter is the basic SI unit of area. The Conférence Générale des Poids et Mesures (C.P.G.M.) recommends that all areas be measured in terms of square meters rather than, say, hectares or square kilometers. A square meter, of course, does not have to be square: an area 4 meters long and 25 cm wide is still a square meter.

acre (ac)

an imperial unit of land area equivalent to 4,840 square yards or 4046.8564 square meters. Historically the acre was a much less precise unit of measurement, being traditionally defined as the area of land a man and an ox could plow in one day. For this reason, it was originally not a square measurement but a long strip, this being a more time-efficient area to plow as it required few turnings of the plow itself. It was defined as the current area of 4,840 square yards by the British Weights and Measures Act of 1878. The U.S. survey acre is fractionally smaller than the international acre, measuring 4,046.8726 square meters.

rood (ro)

an imperial unit of land area equivalent to a quarter of an acre, or approximately 0.1012 hectares. Originally, a rood was an area of land equivalent to 40 rods long by one rod wide. The rood has now all but fallen out of usage.

hide

a non-specific unit of land area. Formerly used across Britain, a hide was deemed to be the area of land that a family and its dependants needed to support themselves. Of course, the amount of land required by a family depended not least on the size of the family, so the hide was generally taken to be anything between 60 and 120 acres. Over time the hide also came to be a measure of tax liability. The Anglo-Saxon Chronicle records that in C.E. 1008 the king (Aethelred II) demanded that every 300 hides should produce one warship and every 8 hides a helmet and a coat of chain mail. Interestingly, no allowance was made for the difference in acreage of each hide.

❧ hundred

historically in Britain, a subdivision of a county or shire which came to be used as a unit of land area. Although the etymology of the word "hundred" in this sense is uncertain, it is thought that it might refer to an area equivalent to 100 hides of land. Another theory relates to the Teutonic invasion of England in the fifth and sixth centuries. The invading armies were divided into forces of 100 men, and it is thought that each "hundred" of men was given a particular area of land when it had been conquered.

Alfred the Great is believed to have been the first king of England to divide shires into hundreds.

riding

an ancient British subdivision of land. Derived from the Old English word *trithing*. which was itself a derivation of the Old Norse *thrithjungr* meaning "third part." the riding was traditionally one of three subdivisions of the county. The term still exists in Yorkshire. which is divided into East Riding. North Riding and West Riding. Ridings were common in Canada in the 19th century: wherever the population size justified it. regions were separate electoral districts. These ridings became more numerous throughout the 20th century as the population increased and the term is still in use today. British counties. e.g.. Gloucestershire. were sometimes divided into four parts. and these were known as farthings. Although the term has now fallen out of common usage. J.R.R. Tolkien made an attempt to keep it in the language by dividing the fictional Shire in *The Lord of the Rings* into four farthings.

county

a subdivision of a country. Originally. a county was that area of Britain under the jurisdiction of a count or earl. Counties are now territorial subdivisions which form a principal unit of local administration. In the U.S.. states are divided into counties for administrative and political reasons. In the U.K. although counties are administrative subdivisions of land. they also traditionally claim populations of distinct characteristics. Originally. British counties were divided into **ridings**.

estate

an area of land or property. The word estate as a land area has various shades of meaning. It is traditionally the grounds surrounding a large country house and owned by the proprietor. Such an estate may include other inhabited properties. or it may just be an area of woodland or other countryside. More recently. an estate has come to be understood to be an area on which similarly designed or featured houses have been built. An estate is also the net total of a person's property and obligations: it may include an area of land. but will also include any other personal possessions or assets.

section

a unit of land area principally used in U.S. and Canadian land surveying. A section is normally equivalent to 1 square mile, or 640 acres. although sections are sometimes slightly smaller or larger to compensate for the curvature of the earth. A **township** consists of 36 sections. and the odd-sized sections are generally placed at the western or eastern edges of a township. Sections are divided into halves and quarters. with the smallest subdivision in common usage being the quarter quarter section. which is equivalent to 40 acres.

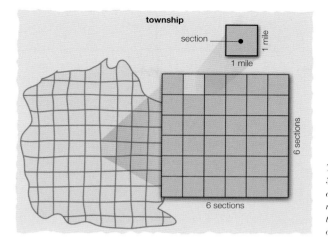

Townships consist of 36 sections, and sections are equivalent to 1 square mile, with some exceptions to allow for the Earth's curvature.

47

☙ township

a unit of land area principally used in U.S. and Canadian land surveying. A township is made up of 36 sections and is equivalent to 36 square miles or 23,040 acres. The township was first defined as a unit of land area by an act of 1785 which stated that its two sides must run north–south, and that the other two sides must be at right angles, i.e., that the township should be square or rectangular. This did not apply if natural features such as rivers made it impractical, or if the township would then cover a Native American reservation.

square mile (sq mi)

an imperial unit of land area. A square mile is the area enclosed by a square whose sides are each 1 mile in length, but it need not be square itself: an area 4 miles long by a quarter of a mile wide is still a square mile. It is equivalent to 640 acres or approximately 2.5 square kilometers. The City of London is traditionally referred to as the Square Mile, but this is only an approximation.

square inch (sq in)

an imperial unit of area. A square inch is the area enclosed by a square whose sides are each 1 inch in length, but it need not be square itself: an area 4 inches long by $^{1}/_{4}$ of an inch wide is still a square inch. It is equivalent to 6.4516 square centimeters, and there are 144 square inches in a square foot.

square rod

an imperial unit of land area, also known as a square perch or a square pole. It is equivalent to approximately 272 square feet or approximately 25 square meters, and only really exists as a unit in North America, though the same measurement is known in Pakistan as the *marla*.

virgate

a non-specific unit of land area. A virgate was historically used in Britain to mean a quarter of a hide. As the hide was a varying unit of area. so the virgate could be anything between 15 and 30 acres.

hacienda

a unit of land measurement deriving from Spanish territories in South America. A *hacienda* was traditionally an area of land awarded to Spanish nobles in Mexico. Argentina and other areas of South America. It is officially calculated as being 89.6 square kilometers. but in fact the area represented by the original *haciendas* varied enormously.

marla

a Pakistani unit of land area. Under British governance of Pakistan the *marla* became a standard measurement equivalent to the **square rod**. or approximately 25 square meters. The *kanal* is the basic unit of land measurement in Pakistan. and is equal to 20 *marlas*.

actus quadratus

a Roman unit of land area. The *actus quadratus* was the basic Roman unit of land measurement. and was a square of 120 by 120 Roman feet. or *pes*. This made it 14.400 square *pes*. which approximates to 13.500 square feet today.

jugerum

a Roman unit of land area equivalent to 2 *acti quadrati*. This makes it equal to 28.800 square *pes*. or approximately 27.000 square feet today. A *jugerum* was thought to be the area of land a yoke of two oxen could plow in a single day.

heredium

a Roman unit of land area equivalent to 2 *jugera*. 57.600 square *pes* or approximately 54.000 square feet today.

aroura

an Egyptian unit of land area. The *aroura* was an area equivalent to 1.000 cubits by 1.000 cubits. This is roughly equivalent to 2.750 square meters. or $2/3$ of an acre.

arpent

the principal unit of land area in France between the 16th and 18th centuries. It was defined as being 100 square *perches*. but this was not a fixed measurement as the *perche* varied according to the region of France. Thus. the *arpent de Paris* was equivalent to approximately 3.420 square meters—this was the most common arpent: the *arpent commune* was equivalent to approximately 4.220 square meters: and the *arpent d'ordonnance* (also known as the *arpent des eaux et forêts* or the *grand arpent*) was equivalent to approximately 5,100 square meters. The *arpent* also exists in Canada, as a result of the French influence there. The Canadian *arpent* derives from the *arpent de Paris*, and so is also equivalent to approximately 3,420 square meters.

ching

a Chinese unit of land area, equivalent to approximately 13.3 square meters or approximately 143 square feet.

ch'ing

a Chinese unit of land area. The *ch'ing* is much larger than the *ching*, being approximately 1 hectare or 2.47 acres.

mu

a Chinese unit of land area. Throughout China's history, the definition of a *mu* has varied wildly, having been as small as 192 square meters during the early Zhou dynasty and as large as 840 square meters during the Yuan dynasty. In 1959, the *mu* was standardized as a metric unit equivalent to 666 2/3 square meters. Related to the *mu* is the *gongmu*, which is a unit equivalent to 100 square meters. The *mu* and the *gongmu* are both used to measure the area of agricultural land. The Taiwanese also use the *mu* as a measurement, but in their system the *mu* is made up of 30 *p'ing* and is equal to approximately 99 square meters.

chüo

a Chinese unit of land area, equivalent to 1,815 square feet or approximately 168.6 square meters.

cong

a Vietnamese unit of land area equivalent to 1,000 square meters or 1,196 square yards.

feddan

a unit of land area used in Egypt and the Sudan. It was historically used throughout North Africa and the Middle East, and is approximately equal to 4,200 square meters. When Egypt adopted the metric system, the feddan was the only one of the old units that remained legally valid. A smaller Syrian *feddan* also exists, but it is not a precise measurement, varying in size between 2,295 and 3,443 square meters.

morgen

a traditional unit of land area used historically throughout northern Europe. In Germany it was for years the most important unit of land measurement, and it still survives in some European areas. Deriving from the German word for morning, a morgen was the amount of land a yoke of oxen could plow in a morning (c.f. **acre** and **jugerum**). As this is of necessity an imprecise measure, the morgen became accepted as different areas in different countries.

In Scandinavia and northern Germany, the morgen became accepted as 0.25 hectares, or approximately 0.63 acres; in Austria and southern Germany, it was more than twice this size, being 0.5755 hectares or 1.422 acres; in the Netherlands, including Dutch colonies, the morgen was 0.85 hectares or about 2.1 acres.

vergee

a unit of land area traditionally used in the British Channel Islands. The size of the *vergee* varied from one parish to another, but in Jersey it was generally around 0.44 acres or 0.178 hectares, whereas in Guernsey it was generally around 0.4 acres or 0.162 hectares. The word *vergee* derives from an old Norman word meaning "orchard."

chomer

an ancient Hebrew unit both of land area and capacity. Translated variously in modern English versions of the Bible as "homer" and "measure," as a capacity the *chomer* was equivalent to about 230 liters, and as land area was deemed to be the amount of land a *chomer* of seed would plant—approximately 2.4 hectares or 6 acres.

rai

a Thai unit of land area. The word *rai* means "field" (an upland field rather than a paddy field). It is an ancient unit of measurement, which is now accepted to be equal to 1,600 square meters or about 0.4 acres. The rai is divided into 4 *ngan*, and it is 1 square *sen*, a *sen* being a measurement of 40 meters. In northern Thailand this unit of measurement is called the *hai*, and in Laos it is the *lai*.

❧ block

a non-specific unit of land area, principally in the U.S. and Canada. Most North American cities form regular street grids. A block is the area of land enclosed by four intersecting streets or, colloquially, the distance from one street to the next parallel street. Distances between streets vary a great deal from city to city, and are generally anywhere between 80 meters and 160 meters. In some cities, such as New York, the streets running in one direction are closer together than the streets running perpendicular to them, giving rise to the terms "long block" and "short block."

Where a block is rectangular, rather than square, the distances between parallel streets are referred to as "long blocks" and "short blocks."

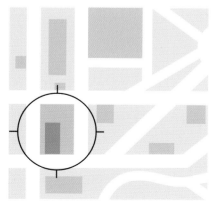

zip code/postal code

alphanumeric systems of dividing land areas into postal regions. ZIP is an acronym for Zone Improvement Plan, and consists of five digits. An extended zip code, called ZIP+4, adds a further four digits to the standard zip code and allows an address to be pinpointed precisely. The first number in the standard zip code represents a particular group of states; the second number represents a region within that group; and the final three numbers define a more specific area within that region. The postal code is a sequence of letters and numbers. The first letter or letters represent a region; the following one or two numbers represent a district within that region; and the final group of letters and numbers represents a more specific area. An individual postal code can represent a street, part of a street or even a single building.

census tract

a subdivision of a particular area for the purposes of a census. It is also referred to as a census area or a census district. Census tracts are designed to be homogenous divisions with regards to the characteristics of a population. The boundaries of census tracts often follow natural geographical features, but they need not necessarily do so. Nor need they be of a particular size with regard to area.

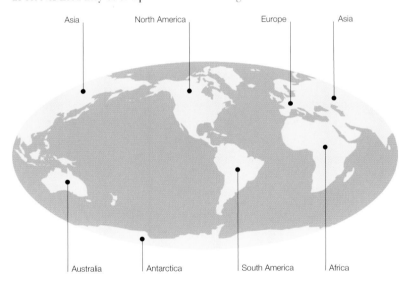

Asia North America Europe Asia

Australia Antarctica South America Africa

❧ continent

any one of the world's expanses of land. The word "continent" derives from the Latin *terra continens*, meaning "continuing tract of land"; as such it is a non-specific unit of measurement, and the number of continents into which the world is divided is not universally agreed. However, it is generally accepted that there are up to seven: Europe, Asia, Africa, North America, South America, Australia and Antarctica.

The world is generally divided into seven continents, though Asia and Europe form one continuous tract of land and Australia is sometimes viewed as part of the Asian land mass.

payload

the capacity of a commercial aircraft for carrying revenue load
(that is, its freight, mail, passengers and baggage). The total available capacity of an aircraft, measured in tonnes, is known as the
payload capacity; the amount of revenue load carried, again in
tonnes, is the payload carried.

ton-mile (tonne-kilometer)

a unit used in commercial aviation for calculating the costs of
moving revenue loads, relating to moving 1 ton of load 1 mile (or
1 tonne of load 1 kilometer). Ton-miles (or tonne-kilometers)
available refers to the **payload** capacity multiplied by the distance
of each stage of a flight. Ton-miles used refers to the payload carried times the distance of each flight stage.

tonnage

in shipping, the measurement of capacity of a vessel for purposes
of registration, assessing tolls, etc. Despite the nomenclature,
tonnage is a measure of volume rather than weight, and usually
refers to gross tonnage: the capacity available for cargo, stores, fuel,
passengers and crew. Until the International Convention on
Tonnage Measurement of Ships in 1969, tonnage was normally expressed in gross registered tons (1 grt = 100 cubic feet), but
nowadays gross tonnage (gt) is defined in terms of cubic meters
using the formula:

$$gt = K_1 V$$
where $K_1 = 0.2 + 0.02 \log_{10} V$
and V = space in cubic meters

The net registered tonnage (nrt), the volume of a vessel's space
for cargo, was similarly affected by the 1969 Convention.
Previously it was simply gross tonnage minus the space for engine, crew and officers, but is now calculated by a complicated
formula involving the draft, passenger-carrying capacity and a
coefficient Kc.

A further, even more complex formula is used to determine
compensated gross tonnage (cgt), which also takes into account
the size and type of vessel. Deadweight (dwt), mercifully, is more
easily calculated. This is the weight in U.K. tons or tonnes of the
cargo, stores, fuel, passengers and crew of the vessel carrying its
maximum summer load. Lightweight tonnage (lwt) is the weight
of the vessel expressed in terms of the weight of water it displaces
in tons or tonnes.

❧ Plimsoll line (loadline)

a mark on the side of a ship showing the maximum permissible draft (distance of keel below waterline) in various seasons and waters. Although officially referred to as the loadline, it is also known widely as the Plimsoll line or mark after British politician Samuel Plimsoll (1824–98), who campaigned tirelessly for stricter regulation of sea-going vessels. He eventually succeeded in getting a Merchant Shipping Bill, which included the mandatory use of a loadline, made law in Britain in 1876.

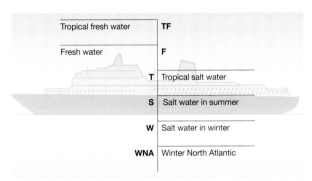

Tropical fresh water	**TF**	
Fresh water	**F**	
	T	Tropical salt water
	S	Salt water in summer
	W	Salt water in winter
	WNA	Winter North Atlantic

Any boat, ship, barge or floating wharf floats at different levels depending on the warmth, consistency or both, of the water. When a ship is fully loaded, the loadline appropriate to the waters in which it is traveling and the season should not be submerged.

shipping ton (freight ton; measurement ton)

a unit of volume (surprisingly not weight) used in shipping, but now obsolete. Originally equal to 40 cubic feet (1.133 cubic meters), the standard measurement ton was an approximation of the space typically occupied by 1 U.K. ton of cargo. Because of the widely differing density of various cargoes, it has also been assigned values of anything from 40 to 100 cubic feet, and its usage varied from trade to trade. Consequently it has fallen into disuse as an accepted standard measure.

register ton

a volumetric unit of internal cargo capacity formerly used in shipping, equal to 100 cubic feet (approx. 2.832 cubic meters). The term derives from its use for specifying the cargo capacity of vessels in registration documents.

twenty-foot equivalent

a unit used in the transport of containerized cargo. In shipping, this is normally quoted in 20-foot equivalent (TEU) or in 40-foot equivalent (FEU), which denotes the length of the container, though other length containers are used for some cargoes. The standard width for these containers is 8 feet (2.4384 meters), and until recently the height was also 8 feet, but these days 40-foot containers are more usually 9 feet 6 inches (2.8956 meters) in height.

❧ tun

a unit of liquid measurement used particularly for wines, spirits and beer in the English-speaking world. It represented the largest size of cask in common usage, and even though sizes varied somewhat, the standard tun contained 200 U.K. gallons (or 252 U.S. gallons—also known as "wine gallons") or 953.88 liters. A tun could be divided into 2 pipes, 4 hogsheads, or 6 tierces (*see* the chart below). The word "tun" is cognate with ton, used in the U.K. and U.S. systems of weights, and its alternatives tonne and tonneau have obvious connections with the SI tonne. Nowadays the beer, wines and spirits wholesale trade is conducted predominantly in liters or hectoliters.

❧ pipe

a unit of liquid measurement used for wines, spirits and beer, equal to half a **tun**, or two **hogsheads** (*see* the chart below). It was also known as a butt, or puncheon. Confusingly, though, the word puncheon also occasionally defines other amounts.

❧ hogshead

a unit of liquid measurement used for wines, spirits and beer, equal to half a pipe, which can be subdivided into two quarters, or four octaves (*see* the chart below).

❧ tierce

a unit of liquid measurement, also sometimes used for dry goods, equal to $1/3$ of a pipe (*see* the chart below). The tierce was used in maritime transport. Although as variable as a **tun** in actual value, the standard tierce was equal to 42 U.S. gallons (158.987 liters); this value survives today in the form of the modern petroleum barrel.

Terms and sizes of containers for beers, wines and spirits. The actual volumes of these containers vary widely according to region, and there are also many different-sized containers specific to particular drinks or areas, such as the butt, aum, leaguer and stück.

1 tun, tonne or tonneau
= 2 pipes

1 pipe, butt or puncheon
= 2 hogsheads = 3 tierces

1 hogshead
= 2 quarters

1 quarter
= 2 octaves

1 tonneau
= 4 barriques

1 muid
= 3 tierçons

1 queue
= 2 pièces

1 pièce
= 2 feuilletés

barrel

a unit of volumetric measurement used historically for trading in liquids such as wines and spirits, and for certain dry goods. Today the barrel is recognized internationally only as a measure of petroleum and 1 barrel is 42 U.S. gallons or 158.987 liters. However, in the English-speaking world the barrel is still used as a unit of measurement for a wide variety of goods, with a confusing number of different values. Beer barrels vary in size between the U.S. and U.K. (1 U.S. barrel is 31 U.S. gallons or 117.35 liters, while 1 U.K. barrel is 26 U.K. gallons or 163.66 liters) and even between U.S. states—in some parts a U.S. barrel may be as much as 42 U.S. gallons. For dry goods, fruit and vegetables the U.S. barrel is 7,056 cubic inches—except for cranberries, where it is 5,826 cubic inches!

As if things weren't complex enough, the barrel is also sometimes used as a unit of weight. The Canadian barrel of cement is 350 lb, whereas in the U.S. it is 280 lb for masonry cement, but 376 lb for Portland cement. Not to be outdone, the U.K. fishing industry still uses the word "barrel" to mean 320 lb (145 kg) of Scottish herring, or 375 lb (170 kg) of whale oil, while the term "continental barrel" means 100 kg.

kilderkin

a variable unit of volumetric measurement for liquid and dry goods, equal to half a barrel in U.S. usage, and consequently dependent on the size of barrel. In the U.K. the term loosely describes a cask of about 16 to 18 U.K. gallons (between 73 and 82 liters).

firkin

a unit of volumetric measurement for liquid and dry goods, equal to half a kilderkin. The standard size is taken to be 9 U.K. gallons (40.9137 liters). In the U.S., the firkin is also used as a measurement of weight, 56 lb (25.401 kg), probably the approximate or average weight of goods contained in a firkin.

rundlet

a unit of liquid measurement equal to 18 U.S. gallons (68.137 liters). Its name comes from an old word for a small cask containing about that volume.

pin

a unit of liquid measurement used mainly in the U.K. for beer. It is equal to half a U.K. firkin, or $4^1/2$ U.K. gallons (20.457 liters).

polypin

a container for beer used in the U.K., usually holding between 32 and 36 U.K. pints. A metric version exists, which contains 20 liters. Polypin is a registered trademark of Biovision GmbH, makers of the polythene-lined container.

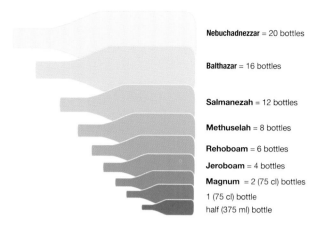

The sizes and names of bottles traditionally used for champagne; in Bordeaux, however, the jeroboam is 5 or 6 bottles, and the methuselah is known as an impériale, *while in the U.K. the jeroboam is 6 bottles, and the rehoboam 8.*

Nebuchadnezzar = 20 bottles

Balthazar = 16 bottles

Salmanezah = 12 bottles

Methuselah = 8 bottles

Rehoboam = 6 bottles

Jeroboam = 4 bottles

Magnum = 2 (75 cl) bottles

1 (75 cl) bottle

half (375 ml) bottle

❧ bottle sizes

bottle sizes for wine range from the "half bottle" at 375 ml to the Nebuchadnezzar at 15 liters. The larger sizes are usually reserved for sparkling wines, particularly champagne. Wines are normally retailed in 75 cl (or sometimes liter) bottles, spirits in 70 cl (sometimes 75 cl or 1 liter), and by the case of 12 bottles. Cheap wine is sometimes sold in liter bottles. Beer and cider commonly retail in bottles of 275, 330, and 500 ml and 1, 1.5, 2 and 3 liter sizes, and in the U.K. draft beer and cider are sold in pubs in pint or half pint measures.

amphora

a container for wine or oil used in ancient Greece and Rome. As with the barrel, the name of the container could also be used for a specific size—around 25 liters in the late Roman period—and the two should not be confused, as the container came in many sizes. The amphora measure was divided into three *modii,* or two *urnae,* and was used for both liquid and dry goods.

modius

a unit of volume used in ancient Rome. It was the equivalent of $^{1}/_{3}$ of an amphora, roughly 8.33 liters.

hectoliter

a unit of volume in the metric system, used widely in the wholesale trading of liquids. It has largely replaced the traditional units such as the barrel, tun, etc., in most trades except petroleum, although there is some resistance to its adoption in some parts of the world, including the U.S., which still use imperial measures.

keg

a container, usually a small cask, for various goods, varying in size and definition accordingly. In the wine trade a keg traditionally

measured 12 U.S. gallons (about 45.52 liters), or half a barrel of
beer, with correspondingly different values in the U.K. and U.S. In
the fishing industry a keg represented 60 herring. As a measure of
nails it was equal to 100 lb (45.359 kg).

cade

a cask used for containing fish. For herring, this was equivalent to
720 fish, or 12 kegs, but the measure has become obsolete.

cran

a unit of volume, equivalent to the contents of a barrel, used in
the U.K. fishing industry. Like the barrel, the cran was originally
a variable measure, but a standard of 37^1/$_2$ U.K. gallons (170.474
liters) became the norm until it was officially abandoned in
1980. The term, believed to be an old Gaelic word meaning a
measure of something, is also used loosely by fishermen to refer
to a hundredweight.

last

an obsolete unit, of variable size, in the U.S. of weight, in the U.K.
of volume. Its U.S. value of around 4,000 lb held for most dry
goods except wool, for which it was equivalent to twelve 364-
pound sacks; in the U.K. it was equal to 80 bushels when referring
to grain, but when referring to fish was the equivalent of 12
barrels—in practical terms, 13,200 herring or 12,000 mackerel.

bushel

a unit of volume with several different values, though now largely
obsolete. In the U.S. it is a unit of dry measurement, also known
as a Winchester bushel, equivalent to 4 pecks, or about 1.2445
cubic feet (35.239 liters). In the U.K. the bushel is a liquid measure, equivalent to 8 U.K. gallons (36.369 liters). In commercial
trading throughout the English-speaking world, agricultural produce such as grain was measured by weight in bushels of various
sizes according to the commodity and the country. The bushel
was later standardized as 60 lb (27.216 kg), and known as the
international corn bushel, but its use has now been superseded by
metric measures.

peck

a unit of volume, a subdivision of the bushel, used primarily for
measurement of dry goods such as grains and fruit. A peck is the
equivalent of quarter of a bushel, or 2 gallons, which in the U.S.
equates to 8.80975 liters and in the U.K. to 9.09225 liters.

sheaf

a traditional (and approximate) unit for measuring grains such as
wheat and barley still on the stalk. A sheaf is a bunch of stalks
roughly 30–36 inches (75–90 cm) in circumference.

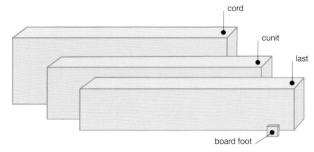

A standard cord is considered a 4′ × 4′× 8′stack of wood, including the bark and airspace. Many of the traditional units of measurement in forestry and the timber trade are still used, particularly in North America, but are gradually being replaced with metric units.

cord

a unit of volume used in forestry to measure stacked roundwood, still widely used in the U.S. It is defined as the volume of a stack of logs 8 feet long, 4 feet wide and 4 feet high, and thus is equal to 128 cubic feet (3.625 cubic meters). The term is also used in the U.S. as a unit of weight of pulpwood, the weight of a volumetric cord of wood, which necessarily varies according to the density of the variety of timber. A derived unit, the cord foot, is $1/8$ of a cord, or 16 cubic feet (0.4531 cubic meters).

Riga last

a unit of volume used in forestry in the U.K., named after the city of Riga in Latvia, a major timber port. When used for square-sawn timber it denotes 80 cubic feet (2.265 cubic meters), while for roundwood it measures 65 cubic feet (1.841 cubic meters), reflecting the actual volume of wood.

board foot

a traditional unit of volume for timber, also known as foot board measure (fbm), board measure, or super foot. One board foot describes the volume of a board 1 foot wide, 1 foot long and 1 inch thick, therefore equivalent to $1/12$ of a cubic foot (0.00283 cubic meters).

hoppus foot

a traditional unit of volume for timber used in U.K. forestry, named after Edward Hoppus, who devised the measure in the 18th century. He advocated that to determine the volume of wood in a log of length L and circumference, or girth, G (in feet), the following formula should be used:

$$L (G/4)^2$$

This gives a result slightly smaller than the actual volume of the log, but a close approximation to the volume of usable timber for practical purposes. The unit of measurement deriving from this formula is the hoppus foot, equal to 1.273 cubic feet (0.0361 cubic meters). Related units still in use in some parts of the world are the hoppus ton, equivalent to 50 hoppus feet (1.8027 cubic meters), and the hoppus board foot, equivalent to $1/12$ of a hoppus foot.

standard

a traditional unit of volume for stacked timber used in forestry. There were three such standards in common use: the St. Petersburg or Petrograd standard (165 cubic feet), the Göteborg or Gothenburg standard (180 cubic feet), and the English standard (270 cubic feet).

cunit

a traditional unit of volume for timber used in forestry. It represents the volume of usable wood. i.e.. excluding the bark and gaps between logs, and one cunit is the equivalent of 100 cubic feet (2.8317 cubic meters) of solid wood.

square

a traditional unit of area used in measuring timber, the equivalent of 100 square feet (9.2903 square meters). The square was used primarily for describing the surface area of finished boards and planks.

rick

a traditional unit of volume for timber. equivalent to $1/3$ of a cord. It was originally defined as the volume of a stack of logs 16 inches long. 8 feet wide and 4 feet high, or 42.667 cubic feet (1.208 cubic meters).

face-cord

a traditional unit of volume for timber. also called a rick.

ring

a unit of quantity used by coopers for boards and staves in barrel-making. The term comes from the metal ring used in transportation to hold the staves together in bundles of 240, and thus represents a quantity of 240 staves or boards. It was divided into four shocks of 60 boards.

bcf

a huge unit of volume. one billion cubic feet. Because of the difference between U.S. and U.K. usage. it represents respectively either one thousand million cubic feet (1.000 mcf) or one million million cubic feet (1.000.000 mcf).

quad

an obsolete unit of heat energy. equal to one U.S. quadrillion (or. in the U.K.. one billiard) British thermal units (10^{15} Btu). equivalent to approximately 1.055 exajoules.

Q unit

a huge unit of heat energy. one thousand quads, or one U.S. quintillion (in the U.K. one trillion) British thermal units (10^{18} Btu). equivalent to approximately 1.055 zettajoules.

❧ apgar score

a means of measuring the health of a baby immediately after it is delivered. Ranging between 0 and 10, the score is derived from observations of five criteria: heart rate, respiratory effort, muscle tone, reflex irritability and skin color. The baby is given a score between 0 and 2 for each criterion at 1 and 5 minutes after birth. Each child, therefore, has two apgar scores. Below 3 is considered critical and the baby may require medical attention; a score of above 7 is considered normal.

The apgar score, named for Virginia Apgar, who devised it in 1952.

sign	score = 0	score = 1	score = 2
heart rate	absent	below 100	above 100
respiratory effort	absent	weak, irregular, or gasping	good, crying
muscle tone	flaccid	some flexion of extremities	well flexed, or active movements of extremities
reflex irritability	no response	grimace/weak cry	good cry
skin color	blue all over, or pale	acrocyanosis	pink all over

body mass index (B.M.I.)

a ratio of a person's weight in kilograms to the square of their height in meters, and a loose indication of good health—or otherwise. The body mass index is a means of calculating whether an individual is underweight, overweight or the correct weight for their height. Generally speaking, a body mass index of less than 18.5 indicates that a person is underweight, whereas a body mass index of more than 25 indicates that a person is overweight. These calculations are only an approximate guide to health, however: the recommended B.M.I.s alter with age, and do not take into account the amount of body fat—a person with a high percentage of muscle (such as a sports man or woman, for instance) may have the same B.M.I. as an obese person, but will not necessarily be overweight.

blood pressure (mmHg)

a measure of the pressure of the blood in the large arteries. Blood pressure is measured in terms of two numbers: the systolic blood pressure (normally between 100 and 135 mmHg in healthy humans) and the diastolic blood pressure (normally between 50 and 90 mmHg in healthy humans). Thus, a blood pressure of 120/80 would indicate a systolic pressure of 120 mmHg and a diastolic pressure of 80 mmHg.

sphygmomanometer

a means of measuring **blood pressure**. An inflatable cuff is placed around the upper arm, and the cuff inflated. The air is then slowly let out of the cuff, reducing the pressure on the arm. When the cuff is full of air, it measures the systolic blood pressure. When the air has been released, it measures the diastolic blood pressure.

blood count

a means of measuring the number of corpuscles in a given volume of blood. Blood contains a large number of different particles, the quantities of which can be a good indication of health. Blood counts are taken either electronically or manually. Electronic blood counts give a more precise measure of the number of particles, but can be less accurate in that they sometimes recognize certain particles incorrectly. Manual blood counts are more subject to errors of measurement, but tend to recognize the particles more accurately.

phagocytic activity

the process of cells engulfing other particles. Phagocytes are cells that can perform phagocytic activity, and include neutrophils and monocytes, types of white blood cell. In animals phagocytic activity is an important function of the immune system.

clotting time

the time taken for blood to clot. The clotting time of blood is a measure of the efficiency of clotting, and by extension of the general health of the blood. The clotting time of the blood of a healthy human can be anywhere between 5 and 15 minutes. If the clotting time is shorter or longer than this, it may indicate one of a number of conditions, including anemia and thrombocytopenia.

blood group

a means of describing the characteristics of human blood. The most common descriptions of blood groups are those of the ABO system. This defines the blood group (A, B or O) according to what type of antigen (A or B) the red blood cells carry on their surface, and what type of antibodies they produce. Certain blood groups are compatible with others, whereas certain blood groups are incompatible.

cardiotachometry

a means of measuring heart rate over a long period of time. Cardiotachometry is undertaken using a cardiotachometer. These instruments measure heart rate in one of two ways. Either they measure the number of beats that occur over a period of time, and divide that number by the time; or they measure the time between successive beats, and work out the heart rate by finding the reciprocal of that time. Many cardiotachometers are set to trigger an alarm if the heart rate falls below a certain level.

❧ electrocardiogram (E.C.G.)

a record of a person's heart rate and other cardiovascular functions, produced by means of electrocardiography. Electrocardiography records the electric voltage in the heart and produces a continuous strip graph. This graph can highlight a number of the symptoms of heart disease, among them abnormal heart rates, electrolyte imbalances, artery blockages and stress levels. In popular culture, an E.C.G. with a flat line indicates that death has occurred. In fact, a "flatline"—technically known as a systole—indicates cardiac arrest with a very bad prognosis, but not necessarily death.

An electrocardiogram showing the waves (P, T and U) and timed intervals used by cardiologists to determine ventricular depolarization and repolarization of the beating heart.

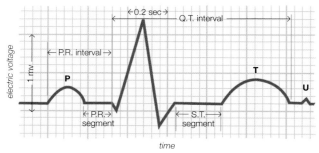

resting heart rate

the number of contractions of the heart in 1 minute when a person is resting. The average resting heart rate is 70. A resting heart rate of less than 60 is a symptom of bradycardia, although it is not necessarily a cause for concern unless it is accompanied by other symptoms of the condition. Very fit people, such as sports men and women, often have a resting heart rate of below 60. A resting heart rate of above 100 is a symptom of a condition called tachycardia.

metabolic rate

the number of calories the human body burns. The basal metabolic rate is the metabolic rate of the body at rest, or the number of calories the body requires in order to maintain its basic functions. Basal metabolic rate differs from person to person according to age, weight, height, diet, fitness and other factors. Metabolic rate increases as a result of exercise, and can be affected by certain conditions such as an overactive thyroid.

❧ electroencephalogram (E.E.G.)

a visual representation of the electrical activity of the brain. An E.E.G. is a neurophysiologic exploration used to assess brain damage, epilepsy, etc., recorded by means of electroencephalography, which involves attaching electrodes to the scalp. E.E.G.s are presented as visual lines either on paper or on an oscilloscope.

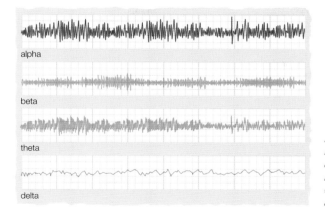

Electroencephalograms record the electrical activity of the brain over time. A flat line where electrical activity = 0 would indicate no brain activity.

❧ brain waves

a measure of the electrical activity of the brain, as represented by an **electroencephalogram**. E.E.G.s measure four different types of brain wave—alpha, beta, delta and theta—each falling within different ranges of frequency. The type of brain wave emitted by the brain alters according to both the state of consciousness and the age of the subject. Theories to explain sleep may rely on E.E.G. patterns recorded on volunteers during sleep sessions.

circadian

a description of a physiological process that naturally occurs in 24-hour cycles, regardless of the presence of natural light fluctuations. Circadian processes are controlled by the body's inbuilt "clock" and include sleep. A gene that controls the body clock has recently been identified in mice. The word circadian derives from the pseudo-Latin *circa diem*, or "around a day."

I.Q.

a measure of a person's cognitive abilities. I.Q. stands for intelligence quotient. A person's I.Q. is measured according to a series of standardized tests, and is normalized so that the average I.Q. of sets of people of certain ages is 100. An I.Q. of more than 100, therefore, indicates that the person's intelligence quotient is above average compared to people of the same age; an I.Q. of less than 100 indicates that it is below average.

twenty-twenty vision

a measure of vision acuity which indicates that a person's vision is perfectly normal, as measured by Snellen's chart. Snellen's chart was devised by the Dutch ophthalmologist Hermann Snellen (1834–1908). In order to score 20/20, an individual must be able to read the 20th line of Snellen's chart at a distance of 20 feet.

optometry

the process and occupation of measuring eyesight. Among other things, optometrists measure the quality of a person's eyesight and prescribe corrective lenses, in the form of spectacles or contact lenses which modify the focal length of the pupil in order to correct long- or short-sightedness.

lung capacity

a measurement of the volume of air that can be absorbed by the lungs during the deepest possible inhalation. Also known as vital capacity (as opposed to tidal volume—the amount of air absorbed by the lungs of a person at rest breathing normally), lung capacity can be measured by capturing the air exhaled after a deepest possible inhalation. It may also be approximated in cubic centimeters by multiplying the surface area of the body by 2,500.

✥ peak flow

a measure of the maximum rate of air breathed out during forced expiration. Peak flow is measured using a peak-flow meter, a short calibrated tube with a mouthpiece. Peak-flow meters are used by people with respiratory conditions such as asthma, and the peak flow can be used as an indication of the severity of the condition or of an oncoming attack.

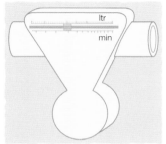

Peak-flow meters are commonly used by asthmatics, but they may be used by people with other respiratory disorders.

respirometer

an instrument used to measure the rate of respiration of an organism. Respirometers measure the intake of oxygen and the output of carbon dioxide.

Breathalyzer

an instrument used to measure the amount of alcohol in a person's breath. As blood passes through the lungs, the air causes a proportion of any alcohol in the bloodstream to evaporate. The alcohol content of the breath, therefore, is directly proportional to the alcohol content of the blood, which is why Breathalyzers give a reliable indication of the alcohol level of the blood. The ratio of blood alcohol to breath alcohol is generally around 2,100:1, though this can change from person to person.

urinalysis

a series of tests used to measure certain characteristics and contents of urine. Urinalysis can establish, among other things: pH of the urine, glucose level and protein level. Because urination is a means of expelling toxins from the body, urinalysis can give a useful indication of certain conditions, including kidney disease and diabetes. It can also be used to detect the use of illicit substances such as recreational and performance-enhancing drugs.

Glasgow coma scale (G.C.S.)

a means of measuring a patient's response to head trauma. The Glasgow coma scale, or score, is calculated by awarding a score to each of three observations: eye response, verbal response and motor response. Each observation is given a score between 1 and 5, and the total of the three scores produces the G.C.S. A score of 3 (the lowest score) indicates that a person is in a deep coma; a score of 15 (the highest score) indicates that a person is fully awake.

paraplegia

paralysis of the lower body. Levels of paralysis are measured according to the proportion of the body paralyzed. Quadriplegia is a form of paralysis affecting all four limbs, whereas hemiplegia is a form of paralysis affecting one side of the body.

sun protection factor (S.P.F.)

a measure of the effectiveness of a sunscreen against the effects of ultraviolet type B radiation which causes sunburn. A sunscreen of S.P.F. 10 means that the user can stay in the sun for 10 times longer than they would be able to without it, although this figure is affected by other factors such as the user's skin type and the strength of the sun. S.P.F. ratings do not indicate the sunscreen's ability to filter ultraviolet type A radiation, which can also have damaging effects on the skin.

burns, degrees of

a measure of burn severity. There are commonly three degrees of burn: first degree, second degree and third degree. The symptoms of a first-degree burn are redness and soreness; a second-degree burn also has some level of blistering; and a third-degree burn includes some charring of the skin. Burns that affect the tissue beneath the skin are sometimes referred to as fourth-degree burns.

computed tomography

a process used to produce a detailed image of a cross-section of the body. Computed tomography uses a series of X-rays taken around a particular axis of rotation. The process was first developed by Godfrey Hounsfield (1919–2004) (**Hounsfield scale**), and images thus produced are known commonly known as CAT scans.

⁊ ultrasound scanning

a means of producing a two- or three-dimensional image of internal organs using high-frequency sound waves. Ultrasound is very good at imaging muscle and soft tissue, and has the advantage that it can produce real-time, moving images on a screen; it cannot, however, penetrate bones. Ultrasound is used routinely on pregnant women to check the health of the fetus at various stages of its development. It is also used widely in other areas of medicine.

Most pregnant women undergo an ultrasound scan at various stages of their pregnancy. The scans produce moving images from which still images can be taken, as shown here.

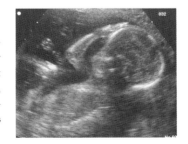

magnetic resonance imaging (M.R.I.)

a means of producing a two-dimensional image of internal organs using magnetism and sound waves. M.R.I. scans resemble X-ray images, although they are somewhat more detailed and have the advantage of not exposing the body to potentially harmful X-ray radiation. More specialized forms of M.R.I. scans include magnetic resonance spectroscopy, functional M.R.I., diffusion M.R.I. and interventional M.R.I.

positron emission tomography (PET)

a means of producing three-dimensional, color images of substances occurring in the body. Unlike CAT scans and M.R.I. scans which provide images of the anatomy of the body, PET scans are used to produce images of the body's metabolic functions. They are used principally in cardiology, neurology and oncology. PET scans use short-lived radioactive substances which mean that the patient is exposed to a similar level of radiation as would be generated by two X-rays.

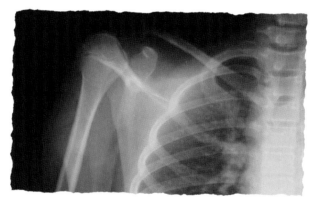

X-rays pass less readily through bone than through soft tissue, which is why radiography can produce useful images of the skeleton.

☙ radiography

the process of creating images on photographic film using X-rays. X-rays have the ability to pass through solid objects, but the strength of the resulting ray is dependent on the density of the object it has passed through. Radiographs can therefore display a two-dimensional image of the internal structure of an organism.

radiology

the medical use of radiation devices. Radiology is a branch of medicine that includes the use of imaging devices such as radiographs, CAT scans, M.R.I. scans and ultrasound scans.

posology

the branch of medicine concerned with dosage. The word posology derives via French from the Greek *posos*, meaning "how much."

L.D.$_{50}$

a measure of toxicity. L.D.$_{50}$ stands for "lethal dose 50 percent," and is defined as the dosage of a substance that kills 50 percent of a trial population. L.D.$_{50}$ is generally expressed in terms of mass of the substance per body mass—e.g., milligrams per kilogram. It is gradually being phased out as a measure of toxicity because it of necessity refers to animals rather than humans. It is problematic because a substance may be more or less toxic to an animal than to a human. Increasing awareness of animal rights has led to the development of alternative means of measurement.

mouse unit (M.U.)

a unit of toxicity. A mouse unit is the amount of a substance that causes death in 50 percent of mice—i.e., it is the L.D.50 for mice. So, the size of a mouse unit varies according to the toxin.

demography

the study of size, structure and distribution of populations. The study of demography includes factors such as incidence of disease, birth rates, death rates, fertility rates, infant mortality rates, life expectancy and reproduction rates. Demographic data can come from a wide range of sources, including birth records, death records and census information.

birth and death rates

the number of births and deaths per population of 1,000 people per year; these are more properly known as the crude birth rate and crude death rate. The crude death rate can be a misleading figure—often the crude death rate of populations in the developing world may be lower than that of populations in developed nations, because life expectancy in developed nations is higher. A more meaningful statistic of mortality is provided by a death rate according to age.

๛ life expectancy

a measure of the average remaining lifetime of an individual in a given population. In populations with a high infant-mortality rate, therefore, life expectancy at birth is considerably different from life expectancy at, say, age five. Colloquially, however, the term life expectancy refers to life expectancy at birth. Average life expectancy has steadily increased with advances in medical science, though certain areas of the world—for example, those affected by the AIDS epidemic—have seen dramatic reductions in life expectancy.

In most parts of the world, life expectancy has gradually increased in line with better public health, especially clean water and improved sewerage systems. This graph shows the average life expectancy of American males and females between 1950 and 2000.

life expectancy (U.S.A.) 1950–2000

80 yrs

70 yrs

60 yrs

female

male

1950   1960   1970   1980   1990   2000

altitude

in meteorology the height of an object above the surface of Earth. above mean sea level or above a constant-pressure surface.

altimeter

an instrument for measuring the altitude of an object. There are several ways of measuring altitude, and so several different types of altimeter: the pressure altimeter measures barometric pressure and compares it with the barometric pressure at sea level in order to calculate an object's altitude: the radio altimeter measures the time taken for a radio signal to travel from a transmitter on the Earth's surface to an object and back to a receiver: and the Global Positioning System (G.P.S.) measures the time taken for radio signals to travel between satellites and a receiver.

lapse rate

the rate of fall of an atmospheric variable (normally temperature) with increase in height.

Weather maps primarily show the position of areas of high or low pressure by means of isobars, and the position and type of weather fronts.

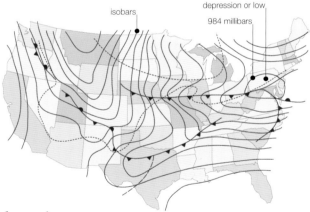

isobars

depression or low

984 millibars

⧫ depression

a region of low barometric pressure. also known as a low or trough. The term depression is usually associated with migratory lows and troughs, upper-level areas of low pressure. and a particular stage in the development of a tropical cyclone.

⧫ bar (b)

a unit of atmospheric pressure equal to 1.01972 kg of force per square centimeter (about 14.50376 lb force per square inch): this is equivalent to a little more than the average pressure of the Earth's atmosphere (1.01325 bar). For practical meteorological purposes. the more common unit is the millibar (mb). A related unit. the barye (ba). is the C.G.S. unit of pressure and equals one microbar. The words derive from the Greek *barys* (weight). from which we also get the terms baromil. barometer. barograph. etc.

millimeter of mercury (mm Hg)

a unit of pressure, equivalent to the pressure of a column of mercury one millimeter high at the Earth's surface. The unit comes from the use of mercury barometers, where the pressure can be read as the height of the column of mercury. In English-speaking countries, the unit was traditionally in inches (e.g.. one bar equals 29.53 inches of mercury), but in meteorology both millimeters and inches of mercury have now been replaced by the bar and millibar. In medicine. however. blood pressure is still measured in mm Hg.

baromil

the unit of length used in calibrating mercury barometers in the C.G.S. system. For a barometer at 45° latitude. at sea level. at 0°C. 1 baromil is equal to an increment of 1 millibar.

ஃ barometer

an instrument for measuring atmospheric pressure. There are two types of barometer: the mercury barometer and the aneroid barometer. The mercury barometer consists of a vertical glass tube containing mercury. closed at the top and immersed in a bath of mercury at the bottom; pressure is read by the height of the column of mercury in this tube. Less accurate. but more portable and compact. is the aneroid barometer. which consists of a vacuum chamber of thin corrugated metal in the form of a bellows. with one end fixed and the other attached to a mechanism for converting the movement of the bellows caused by variations in atmospheric pressure to a pointer on a scale.

thermobarograph

an instrument for recording pressure and temperature. a combined thermograph (recording thermometer) and barograph (recording barometer). The most commonly used thermograph in weather stations consists of a bimetallic spiral attached to a recording pen. and the most common barograph is of the aneroid type. also attached to a recording pen. The pens mark a piece of paper on a revolving drum. thus effectively creating a graph of both temperature and atmospheric pressure against time.

isobar

a line on a weather map connecting places having the same atmospheric pressure at a given time.

isotherm

a line on a weather map connecting places having the same temperature at a given time.

The atmospheric pressure acting on the surface of a bath of mercury is sufficient to support a column of mercury. The space at the top of the calibrated tube is known as the Torricellian vacuum.

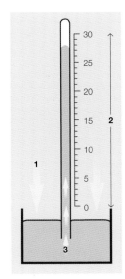

69

1   *Air pressure presses down on the mercury. forcing it up the evacuated glass tube.*
2   *Scale measures height of mercury.*
3   *Covering keeps mercury from spilling. but allows air pressure to influence mercury.*

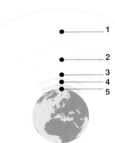

Altitudes of the different
levels of atmosphere are:
1 exosphere 700–2,500 km
2 thermosphere 85–700 km
3 mesosphere 50–85 km
4 stratosphere about
   12–50 km
5 troposphere up to about
   12 km

The ozone layer lies in the
stratosphere, 10–50 km
above sea level. Above this
are the D-Layer (56–83 km),
E- or Kenelly-Heaviside
Layer (83–153 km), and
the F- or Appleton Layer
(153–400 km), which form
the ionosphere.

❧ atmosphere

the sphere of gases that surrounds the surface of Earth. The Earth's atmosphere extends to an altitude of about 2,500 km, and can be divided into different layers: the heterosphere, comprising the exosphere and the thermosphere; and the homosphere, comprising the mesosphere, stratosphere and troposphere. Other divisions include the ozone layer, and the highly ionized D-, E- and F-layers.

wind speed

the speed of movement of air across the Earth's surface, measured in meters per second, or traditionally in miles per hour or knots. Wind **velocity** also includes wind direction, which is given as the direction (as a compass point) from which the wind blows.

Beaufort wind scale

an empirical scale for estimating wind speeds based on observable effects of the wind. It was originally devised by Admiral Sir Francis Beaufort (1774–1857) for use by seamen and was based on the effect of the wind on waves at sea. Later it was developed as a scale for use on land, with corresponding observable effects on land. The internationally recognized scale of 0 to 12 was extended to 17 by the U.S. Weather Bureau in 1955, but the additional force numbers 13 to 17 are generally considered impractical and unnecessary.

wind chill factor

a measurement of the cooling effect of wind speed combined with low temperature on the surface of the human body. Temperatures are perceived as lower than the actual air temperature because of wind, and wind chill is the rate of heat loss from exposed skin caused by this movement of air. Antarctic explorer Paul Siple first used the term in 1939, and subsequently meteorologists devised a wind chill index (W.C.I.) to take into account this phenomenon. The original wind chill index was refined in 2001 to give the wind chill temperature index (W.C.T.I.), using the formula, wind chill = $13.12 + 0.6215T - 11.37(V^{0.16}) + 0.3965T(V^{0.16})$, where T = temperature in °C, and V = wind speed in km/h.

Coriolis force

a hypothetical force used to explain the movement of objects in a rotating system, specifically in meteorology to account for the deviation from the expected movement of bodies due to the rotation of Earth. It can be seen in the apparent drift eastward of winds blowing north from the Equator, due to the comparative difference of speeds of rotation at different latitudes. The Coriolis effect is named after French mathematician Gaspard Gustave de Coriolis (1792–1843), who first described the phenomenon and proposed a fictitious force to compensate for the apparent deviation from Newton's second law of motion.

Fujita tornado scale

an empirical scale for measuring the wind speed of tornadoes based on observable damage caused by them. More correctly called the Fujita–Pearson Scale, it takes its name from American meteorologists Theodore Fujita and Allen Pearson, who developed the scale in 1971. The scale is similar to the Beaufort scale in its use of force numbers, ranging from F0 to F5, to denote increasing wind speeds.

tropical cyclone intensity scale

a scale for describing the severity of revolving storms (including cyclones, hurricanes and typhoons) that originate over the tropical oceans according to their average wind speeds at standard anemometer level (10 meters). The internationally agreed scale of intensity is:

1. tropical depression, with winds up to 17 meters per second
2. tropical storm, with winds of 18–32 meters per second
3. severe tropical cyclone, hurricane or typhoon, with winds of 33 meters per second or higher.

Saffir–Simpson hurricane scale

a scale for describing the severity of hurricanes based on observable damage caused by them. It is named after American structural engineer Herbert Saffir, and meteorologist Bob Simpson, director of the National Hurricane Center, who devised the scale in 1969.

drought severity scale

a scale for measuring the severity of droughts in specific locations at specific times, taking into account the precipitation deficit and the abnormality of the weather. In 1965, American meteorologist W.C. Palmer devised a scale known as the Palmer Drought Severity Index (P.D.S.I.), which was calculated from precipitation and temperature figures, and the available water content of the soil.

☙ tides (neap and spring)

in astronomical terms, tides are the distortions of the surface of a planet caused by gravitational forces acting on it. However, in meteorological terms, tides are specifically the ocean tides on Earth caused mainly by the gravitational force of the moon, and to some extent by that of the sun. This attraction causes a "bulge" in the water of the oceans facing toward the moon, and a corresponding bulge on the opposite side of Earth, which are the areas of high tides. When the sun's gravity reinforces that of the moon (at the time of full and new moons), there are stronger tides known as spring tides. When the sun's gravity is acting at 90° to the moon's (at the time of half moons), there are weaker tides known as neap tides.

At the time of the spring tides, high tides are higher and low tides lower than average; at the time of the neap tides, the difference between high and low tides is smaller than average.

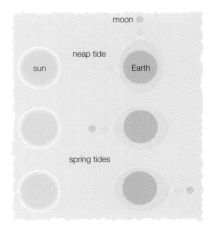

common weather symbols

rain

light   moderate   heavy

shower

light shower   moderate shower

storm

thunderstorm   heavy thunderstorm

drizzle

light   moderate   heavy

freezing rain

light   moderate

snow

light   moderate   heavy

light shower   moderate shower

other

haze   fog

ice crystals

As well as showing barometric pressure, wind speeds and weather fronts, weather maps often indicate various forms of precipitation with graphic symbols.

humidity

the amount of moisture in the air. It is usually (and more correctly) expressed as relative humidity, the ratio of vapor pressure in moist air to the saturation vapor pressure with respect to water at the same temperature, given as a percentage. The relative humidity of a given sample of air is greatly affected by temperature, and conversely the relative humidity affects the apparent temperature. This relationship can be seen in a heat index chart (also called a comfort chart), similar to a wind chill chart, which gives the perceived temperature at various temperatures and relative humidities.

dewpoint

the temperature at which a given sample of moist air becomes saturated and deposits dew when in contact with the ground or a cooler surface, or condenses into water droplets above ground (when it is occasionally called the cloud point).

hygrometer

an instrument for measuring atmospheric humidity. There are several types of hygrometer, including the wet and dry bulb hygrometer (also known as a psychrometer) which consists of two thermometers, one with its bulb surrounded by a damp wick dipping into water, which gives a reading affected by the cooling effect of the relative humidity of the ambient air, the other giving a "true" reading of the temperature. A calculation can then be made of relative humidity by comparison of the two readings. Electrical hygrometers use resistance elements, known as hygristors, which are sensitive to ambient humidity.

rainfall

the amount of moisture falling on the Earth's surface at a given point in a given period, usually expressed in centimeters or inches. Meteorological usage has firmly established the term "rainfall," when "precipitation" would provide a more accurate description, as this term includes snow and hail—and when referring to the results of a traditional rain gauge also may contain small amounts of condensation.

pluviometer

an instrument for measuring rainfall, also known as a rain gauge. The most common type consists of a funnel leading to a narrow calibrated collecting container, enabling accurate readings of rainfall over a given period, usually 24 hours.

hyetograph

a chart or diagram showing the time and amount of precipitation in a given place. The data for hyetographs can be taken from a continuously recording rain gauge such as the Dines tilting siphon.

sonde

a device carried in a balloon, satellite or rocket to send back data such as measurements of temperature, pressure and humidity. A radiosonde transmits this data by radio to a ground receiver, and can send details of the successive levels of the atmosphere as it ascends to the stratosphere.

inversion

a reversal of the expected decrease in temperature with altitude in the atmosphere. An inversion layer is a layer of air that is warmer than the layer below it. Inversions are not uncommon, and occur on clear nights and in anticyclones.

visibility

the maximum distance at which a sufficiently large dark object can be seen against the sky at the horizon in normal daylight, or a moderately intense light can be seen at night. Normally, visibility readings are made at several points around the horizon circle and averaged out to give a single reading for that location.

transmissometer (transmittance meter)

an instrument for measuring visibility, or more precisely the transmission or extinction coefficient of the atmosphere to determine visibility. It is also known as a telephotometer or hazemeter.

cloud base

the lowest altitude at which the air contains a measurable quantity of cloud particles in a given cloud or cloud layer, also known as base of cloud cover. The height of the cloud base above local terrain is known as the cloud height, and the vertical distance from cloud base to the cloud top is known as the thickness or depth of cloud.

❧ cloud types

clouds are classified according to shape, and the height at which they occur, and were given their Latin names by English meteorologist, Luke Howard, in 1803. *Cirrus* means curl of hair, *stratus* means layer, *cumulus* means heap. There are 10 main cloud types, each having its base at one of three levels: high clouds, above 20,000 feet (6,000 m); mid-level clouds, at 6,500 to 20,000 feet (2,000–6,000 m); and low clouds, below 6,500 feet (2,000 m).

Cirrus: detached wispy white clouds of ice crystals.
Cirrocumulus: small round clouds of ice crystals and water droplets, often regular ripple-like formations.
Cirrostratus: continuous white veil of cloud formed of ice crystals.
Altostratus: gray/blue thick sheet of cloud containing water droplets.
Altocumulus: gray or white round clouds, often ripple-like formations.
Cumulonimbus: vertically developing and towering clouds, dark at the base, with anvil shaped white tops.
Cumulus: vertically developing, billowy white "cotton-wool" cloud, often with gray base.
Stratus: sheets/patches of gray, shapeless cloud, often starting as fog.
Stratocumulus: sheets/patches of round gray cloud.
Nimbostratus (not shown): dark, low, shapeless rain clouds that merge in patches.

cirrus · cirrocumulus · cirrostratus · altostratus · altocumulus · cumulonimbus · cumulus · stratus · stratocumulus

minerals and metals

streak

the color of a material in powdered form, and therefore the color that a "streak" drawn with the material across a harder substance will take. Often a useful way of identifying materials that otherwise look similar to the eye. For example, the naturally occurring rocks magnetite (magnetic iron oxide) and chromite (the main source of chromium metal) are both dark gray in appearance, and often occur together, but chromite's streak is brown, while that of magnetite is black. Conversely, crystals of quartz can appear in many colors, but the streak is always white.

permeability

the ability of a substance to allow liquid or gas to flow through. It depends on the size of the substance's pores, and the extent to which they are connected together. An important consideration in mining, since the permeability of the surrounding rock controls the amount of water that will flow into a mine once the workings pass below the local groundwater level.

darcy

a unit of permeability. Defined as the volume (in cubic centimeters) of a liquid with viscosity one centipoise to pass through 1 square centimeter of the substance's surface per second under a pressure gradient of 1 atmosphere per centimeter. Typical sandstone (a very porous rock) has a permeability of about 1 darcy—most rocks have permeabilities of a few tens of millidarcys. On the other hand, a sand filter for a swimming pool will usually have a permeability of tens—even hundreds—of darcys. The SI has no special name for the unit of permeability (though the dimensions are those of area): 1 darcy equals about 10–12 m^2.

assay ton (A.T.)

a measure of the amount of precious metal contained by an ore, equal to the number of troy ounces of pure metal that would be obtained from a ton of ore. In conversion to metric units, the assay ton suffers from the difference between the long (British) ton of 2,240 lb and the short (U.S.) ton of 2,000 lb: the former giving a ratio of 1 gram of metal to 32.7 kg of ore, while the latter is 1 gram of metal to only 29.2 kg ore.

✺ vein

a deposit of metal ore, or other valuable substance, e.g., gemstones. Usually formed in a fissure, fault or similar defect in the local rock structure by materials traveling upward from deeper underground.

Natural diamonds were created deep underground under immense pressures, and are brought to the surface by a special kind of eruption through a volcanic pipe called a "kimberlite."

1 tuff cone
2 crater zone
3 kimberlite
4 diatreme zone
5 dykes
6 root zone

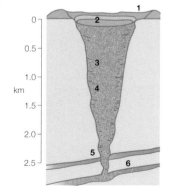

❧ ore grade

the proportion of the desired metal(s) in an ore. usually given as a percentage of the total weight for common metals. Precious metals and others (such as uranium) with low concentrations will often be given in grams per tonne.

cut-off

the lowest ore grade from which it is worth extracting a metal. The level of cut-off is an economic factor in determining whether a particular vein

element	average abundance (ppm)	minimum ore grade (ppm)	factor
gold	.004	0.5	125
molybdenum	2	500	250
tin	2	500	250
lead	12	15000	1250
copper	55	5500	100
zinc	70	5000	700

The average abundance in the Earth's crust and cut-off grades for various metals.

is worth mining. Cut-offs vary enormously from one material to another—the economic cut-off for iron ore is around 55–60 percent. while gold can be mined profitably at less than a gram per tonne in easily accessible open-cast workings. Ore with a grade below the cut-off is rarely mined deliberately. and is usually discarded as waste material.

parts per thousand (ppt)

a unit of proportion. Sometimes used in the form "per mil" (or "per mill"). similar to percent. For some minerals. useful as a grade (1 ppt is 1 kg per metric tonne). although percentages and **parts per million** are more common.

parts per million (ppm)

a unit of proportion. equal to grams per (metric) tonne. Widely used for precious metal ore grades. as well as for pollutants and other chemicals in water.

trace

a very low concentration of a mineral. sometimes defined as anything less than **1 ppm**. Sometimes this will be the main desired substance (for precious metals. for example). but it can also be a convenient (or inconvenient) impurity in a vein. A related use is the "trace elements." which the human body needs in only very small quantities (e.g.. chromium and copper)—but which may be mined in much larger ones (cut-off for copper is around 2 percent).

❧ reserves

the amount of ore. fossil fuel. or other valuable substance contained within a particular deposit or **vein**. Normally quoted as a number

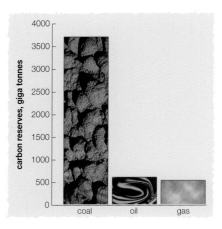

When calculations are made to determine how much energy the (very different) reserves of fossil fuels hold, their different "fuel value" (see heat of combustion) has to be taken into account.

of tonnes (or barrels. etc.) of the ore on the assumption that the **cut-off** remains the same throughout the deposit. For metals. it is sometimes quoted as a number of tonnes of ore with a specified average grade. Mining and oil companies also state their total reserves, simply the sum of the reserves in all their claimed deposits.

sieve number

a measure of the opening size in a mesh sieve. and therefore of the largest particle that can pass through it. The sieve (or mesh) number is the number of holes per unit distance. Modern mm-based numbers are standardized. but British and American usage relating to inches differs—in the U.K.. the sieve number specifies the number of units of actual mesh (including wires) per inch. and the gauge of the wire is therefore part of the sieve specification. U.S. sieve numbers are the size of the hole as a fraction of an inch. and therefore have larger holes for the same sieve number.

slate sizes

roofing slates come in a variety of standard sizes: as with many other traditional materials. these sizes are given names rather than the more modern habit of merely referring to the size itself.

A "wide" slate is 2 inches wider than standard: a "small" one 2 inches shorter—thus a wide Duchess is $24'' \times 14''$ and a small Lady $14'' \times 8''$.

Standard sizes of roofing slates
length by width, in inches:

Duchess	24 x 12
Marchioness	22 x 11
Countess	20 x 10
Viscountess	18 x 9
Lady	16 x 8
Header	14 x 10
Double	12 x 6

The traditional names for roofing slates have been said to be falling into disuse for many years now. but can still be found on the Web sites of numerous firms selling traditional slates.

phi scale

a scale of particle size based on average diameter. intended primarily for sediments. sand and other small objects. A 0ϕ object is (approximately) 1 mm in diameter: the diameter is halved for every step in the scale (so a 2ϕ object is 0.25 mm across). Negative numbers are used for larger objects—a 1-inch diameter object is about -4.7ϕ.

coal/particle sizes

A wide range of sizing schemes exists. with coal being classified differently to other substances in many countries. Different types of cold-fired furnace often work best with specific sizes of coal.

A.S.T.M. grain size index

a measure of the size of the "grains" (or crystals) making up the internal structure of a metal or other substance. The grain size index is N where the number of grains per square inch at 100 times magnification is $2^{(N-1)}$. (So 1 for 1 grain per square inch. 2 for 2. 3 for 4 grains per square inch. etc.)

metal fatigue

a process by which metal objects suffer a gradual loss of strength, and eventually break as a result of fluctuating stresses that always remain below the tensile strength of the object. Continually changing stresses cause slow changes in the structure of the material, allowing a crack to form at a point of stress concentration and then slowly extend until it reaches the **Griffith crack length** and breaks completely. Normally considered only in terms of metals, fatigue damage can occur in almost any material, though the precise mechanisms of failure vary.

phi units*	size	Wentworth size class	sediment/rock name
-8	256mm	boulders	sediment: gravel
-6	64mm	cobbles	
-2	4mm	pebbles	rock rudites: (conglomerates, breccias)
-1	2mm	granules	
0	1mm	very coarse sand	
1	1/2mm	coarse sand	sediment: sand
2	1/4mm	medium sand	rocks: sandstones (arenites, wackes)
3	1/8mm	fine sand	
4	1/16mm	very fine sand	
8	1/256mm	silt	sediment: mud
		clay	rocks: lutites (mudrocks)

*Udden-Wentworth scale

The phi scale table is used by mineralogists and geologists to grade sediments by their particle size.

fatigue limit

the stress below which a material does not suffer any fatigue damage, regardless of the number of times the stress is repeated. Many materials (e.g., aluminum) do not have a fatigue limit, and will eventually fail completely under repeated loadings with even the smallest stress; the magnitude of the stress only changes the number of fatigue cycles that the material can survive. The fatigue strength of a material is a related but different measure, being the stress at which the material will only fail after at least a specified number of cycles (usually between 10,000 and 10 million, depending on the purpose of the object).

ductility

a combined measure of the ease with which a material can be drawn into wires, rods, sheets, etc., and the ability of the material to withstand such treatment. Generally speaking, metals and plastics are ductile, while most other materials are not, but there are exceptions on both sides.

malleability

the ease with which a material can be hammered (or otherwise compressed) into shape, and how much forming of that type it can accept. A highly ductile material will often also be extremely malleable (both processes are forms of plastic deformation, where the shape of the object is permanently changed), but the two do not always go together. Glass, for example, becomes relatively ductile well below its melting point, at temperatures where it will still shatter instantly when hit with a hammer.

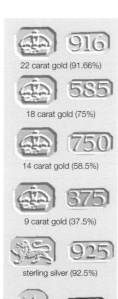

22 carat gold (91.66%)

18 carat gold (75%)

14 carat gold (58.5%)

9 carat gold (37.5%)

sterling silver (92.5%)

Britannia silver (95.8%)

Hallmarks have been used in Britain and France since at least the 14th century. Modern British usage has a crown and a fineness number—before 1975, the karat purity was used for gold. The shape of the "shield" around the number indicates the metal . The lion indicates 92.5 percent silver, while the leopard's head is one of several marks used to show where the metal's purity was checked. Other countries use different systems (the U.S. has no official system at all), often much more complex.

Bullion bars for trading and transport are normally "fine silver," meaning 999 fine.

❧ karat (kt), carat (ct.)

a measure of the purity of gold, being the number of 24 equal parts by weight that are gold. Thus 24 karat gold is pure (as far as possible), 18 karat is 75 percent gold, and so on. "Colored" white and red gold can never reach a purity of more than around 18 karat, since the color is produced by the impurities (copper in red gold, and nickel and platinum in white). Karat (metal purity) and carat (a unit of mass used for gemstones) are used differently in the U.S., though the root is the same, and the two are identical in many places; in the U.K., carat is used for both; in Germany, *Karat*. As a unit of mass, a value of 200 mg was officially adopted for the carat (also known as the metric carat) in 1907.

fineness

a measure of the purity of precious metals, equal to one part in a thousand. Expressed as a number between one and 1,000, so 18 karat gold is 750 fine. Occasionally also given (for extremely pure silver) as a number between zero and one representing the proportion of silver—usually 0.999 (999 fine) or higher. The highest purity in significant production is 0.99999, containing less than 0.001 percent of other metals.

silver grades

a variety of names exist for standard purities of silver. "Coin silver" is 90 percent silver (the other 10 percent is usually copper), and used for making commemorative coins in various countries. "Mexican silver" is usually 95 percent silver and 5 percent copper. The two main grades used for jewelery and decorative items (and commemorative coins, in many countries) are "Sterling silver" (92.5 percent) and "Britannia silver" (95.8 percent).

gold   sterling silver   gold, silver, platinum
(since 1975)

gauge (of sheet metal and wire)

a measure of the thickness of a wire or sheet of metal. In traditional American and British systems in inches, higher numbers indicate thinner material. Negative numbers are not used, with the gauge thicker than 1 being 0, the next either 00 or 2/0, the next 000 or 3/0, and so on. There is, however, no fixed standard for the conversion from inches to gauge number. Metric wire gauges are much simpler, with the number being simply 10 times the thickness in millimeters and therefore increasing for thicker wires.

time and calendar

Planck time

a very small unit of time—probably the shortest unit of time possible within the present laws of physics. It is defined as the time taken by a photon to travel 1 **Planck length** at the speed of light, and is equal to about 1.351×10^{-43} second. Planck time is named after German physicist Max Planck (1858–1947).

second (s; sec)

the base unit of time in the SI system. Originally defined as $^1/_{86,400}$ of a mean solar day, then refined to specify $^1/_{86,400}$ of the mean solar day January 1, 1900, the internationally accepted definition was fixed in 1967 in terms of the frequency of radiation corresponding to the transition between the two hyperfine levels of the ground state of the caesium-133 atom, the duration of 9,192,631,770 periods of this particular radiation being exactly 1 second. From the second, other units of time, such as the minute and hour, can be derived.

metric second

a small unit of time in a proposed "metric time" system. The idea of decimalizing time has been around for centuries, and there have been several proposed versions of this "metric time": most divide the day into 10 "metric hours", each consisting of 100 "metric minutes" of 100 "metric seconds" (also called beats, or blinks). As the day seems to be the base unit of most of these systems, precise definitions depend on what actually constitutes "a day."

minute (min)

a unit of time equal to 60 seconds. Before the adoption of the second as the base unit of the SI system, it was defined in terms of being $^1/_{60}$ of an hour, or $^1/_{1440}$ of a day. The **sidereal** day is divided in this way.

⚓ bell (ship's time)

a unit of time, a division of the **watch**, used especially on board ships: each four-hour watch is divided into eight bells of 30 minutes. At sea, a bell is struck every half hour to mark the division of the watch, and is sounded the same number of times as the number of bells that have occurred since the beginning of that watch.

hour (h; hr)

a unit of time equal to 3,600 seconds, or 60 minutes. Before the adoption of the second as the base unit of the SI system, the hour was thought of, loosely speaking, as $^1/_{24}$ of a day, and the **sidereal** hour is in fact $^1/_{24}$ of a sidereal day. The idea of a 24-hour day stems from the ancient division of the daylight into 12 periods

In horology the term ship's bell refers to a clock which strikes according to a system akin to that used on board ship where a bell is struck manually up to eight times to denote the four-hour "watches" or the periods of duty.

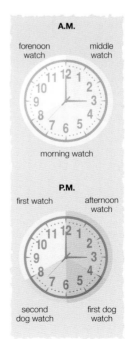

A.M.

forenoon watch — middle watch

morning watch

P.M.

first watch — afternoon watch

second dog watch — first dog watch

(making summer hours longer than winter hours!), and the subsequent division of the nighttime also into 12. With the advent of mechanical clocks, the obvious next step was to have 24 equal hours to the day. The word hour derives from the Greek word *hora* (period), originally meaning a season, but was adopted by the Christian Church for the services held at specific times of each day, thus giving its modern sense of a division of the day.

❧ watch

a unit of time denoting the period of duty of a sentry or ship's crew, mostly, but not always, lasting 4 hours. On ships at sea, the day is divided into seven watches. A ship's bell clock, though, may not necessarily accord with the nautical time since, starting at 12:30 p.m., a ship's bell clock strikes one bell and an extra blow is added at each half hour until 4 p.m. when eight bells are sounded.

day (d; da)

a unit of time traditionally defined as the period taken for one rotation of Earth on its axis in relation to the sun; in everyday usage a period of 24 hours. This definition, however, is not as accurate as it may seem, and more specific ones have evolved. To astronomers, Earth completes one rotation in relation to distant stars in 23 hours 56 minutes 4.09054 seconds (the extra approximately 3 minutes 56 seconds is apparent to us on Earth only because of our orbit around the sun), and they refer to this period as the **sidereal** day; this is divided into 24 sidereal hours of 60 sidereal minutes. Here on Earth, however, the day is dictated by our relationship with the sun, and civil time, as opposed to sidereal time, is reckoned accordingly: the average period occurring between two successive noons (when the sun crosses the meridian) is known as the mean solar day. The adoption of the second as the base unit of time of the SI system has also introduced complications: our present definition of the **second** is independent of the behavior of the planets, so that days calculated as 86,400 seconds (60 × 60 × 24) are not in step with the mean solar day—particularly as this is increasing very gradually day by day—and occasional leap seconds have to be inserted to make up for the anomaly.

week

a unit of time equal to 7 days. The 7-day week has been used for thousands of years, despite its awkward relationship with the 365-day year, probably because of religious and astrological associations. Attempts to decimalize the calendar with the substitution of ten-day weeks (*see* **decade**) were made in revolutionary France and the U.S.S.R., but did not catch on.

uinal

a unit of time in the Mayan calendar, equal to 20 days (*see* p. 87).

Horologists have developed clocks that show not only the time of day but also such astronomical information as the phases of the moon, and astrological information about the position of Earth and the sun in relation to the signs of the zodiac.

Jewish calendar months	Gregorian calendar months	Islamic lunar months	Hindu lunar months
no. of days	no. of days	no. of days	no. of days
Tishri 30	January 31	Muharram 30	Chaitra 30
Marcheshvan 29 or 30	February 28 or 29	Safar 29	Vaisakha 31
Kislev 29 or 30	March 31	Rabi'a I 30	Jyaistha 31
Tebet 29	April 30	Rabi'a II 29	Asadha 31
Shebat 30	May 31	Jumada I 30	Sravana 31
Adar 29 or 30	June 30	Jumada II 29	Bhadrapada 31
Nisan 30	July 31	Rajab 30	Asvina 30
Iyar 29	August 31	Sha'ban 29	Karttika 30
Sivan 30	September 30	Ramadan 30	Margasirsa 30
Tammuz 29	October 31	Shawwal 29	Pausa 30
Ab 30	November 30	Dhul-Qa'da 30	Magha 30
Elui 29	December 31	Dhul-Hijia 29 or 30	Phalguna 30

Note: Ve-Adar (29) is the Jewish intercalary month that features every 3rd, 6th, 8th, 11th, 14th, 17th and 19th year

The Jewish, Gregorian, Islamic and Hindu calendars compared. All the major calendars in use today have some form of intercalary unit to compensate for the awkwardness of the 365.242199 days in a tropical year.

ɣ⬥ lunar month

the period of time between two successive passages of the moon through opposition or conjunction. for example the period between two new moons. It is more properly known as the synodic month. and occasionally called a lunation. It is equal to 29.53059 days.

ɣ⬥ calendar month

a division of the year according to a particular calendar. most usually the **Gregorian calendar**. Because the year is irregularly divided into 12 by the Gregorian calendar. the length of a calendar month can vary from 28 to 31 days. Other calendars. such as the Jewish and Chinese. have variable-length calendar months and years. whereas the Islamic calendar uses lunar months.

trimester

a unit of time equal to $\frac{1}{4}$ of a year. also known as a quarter. In medicine. the term is used to denote the three 14-week stages of the human gestation period. whereas in schools or colleges it means an academic term of about 14 weeks. It derives from the Latin. meaning three months.

semester

a unit of time equal to half a year. or 6 months. The term is principally used in schools and colleges to mean half an academic year. and thus can vary from about 15 to 21 weeks.

tun

a unit of time in the Mayan calendar equal to 18 *uinal* or 360 *kin* (days. see p. 87). so analogous to a year in other calendars.

The Hindu solar (zodiac) calendar. The astrological signs of the zodiac originated in Mesopotamia, but rapidly spread throughout Europe and Asia. The astrological year is divided into 12 months, each assigned its own astrological sign derived from the constellation in which the sun appears.

1 *Maysha (Aries) the Ram*
2 *Vrushabha (Taurus) the Bull*
3 *Mithuna (Gemini) the Twins*
4 *Karka (Cancer) the Crab*
5 *Simha (Leo) the Lion*
6 *Kanya (Virgo) the Maiden*
7 *Tula (Libra) the Scales*
8 *Vrushchika (Scorpio) the Scorpion*
9 *Dhanu (Saggitarius) the Bow*
10 *Makar (Capricorn) the Goat*
11 *Kumbha (Aquarius) the Pot*
12 *Meena (Pisces) the Fish*

minimal common year

the shortest of the six types of year in the Jewish calendar, equal to 353 days. In a minimal common year, the months Marcheshvan, Kislev and Adar all have 29 days, and the intercalary month Ve-Adar does not appear. In a minimal leap year (383 days), Adar increases to 30 days, and the 29-day Ve-Adar is included as the 13th month.

regular common year

one of the six types of year in the Jewish calendar, equal to 354 days. In a regular common year, the months Marcheshvan and Kislev have 29 or 30 days according to the year, and Adar has 29. In a regular leap year (384 days), Adar increases to 30 days, and the 29-day Ve-Adar is included as the 13th month.

full common year

one of the six types of year in the Jewish calendar, equal to 355 days. In a regular common year, the months Marcheshvan and Kislev each have 30 days, and Adar has 29. In a full leap year (385 days), Adar increases to 30 days, and the 29-day Ve-Adar is included as the 13th month.

sidereal year

an astronomical unit of time, the period between two successive passages of the sun as seen from Earth through the same point relative to distant stars. Sidereal time is reckoned by the movement of Earth in relation to distant stars, as opposed to civil time, which is measured relative to the sun: in sidereal terms Earth makes about 366.242 rotations during each revolution around the sun (not 365.242, as one rotation is canceled out by that revolution). So, despite the sidereal day being shorter than the mean solar day, the sidereal year is equal to 365.25636 days, a little longer than the **tropical year**.

ᘍ year (a; y; yr)

in general use, a period of time equal to 365 or 366 days, roughly the period taken for Earth to complete one revolution around the sun. A more precise definition is provided by the term **tropical (or solar) year**. As Earth actually takes about 365.242 days to make its way around the sun, calendars using whole numbers of days soon get wildly out of step with the true situation, and the notion of a 365-day year with a leap year every fourth year was introduced by the **Julian calendar**, and further refined in the **Gregorian calendar**. In calendars of other cultures, the term year can mean anything from about 350 to 385 days.

rat · ox · tiger · rabbit · dragon · snake
horse · sheep · monkey · rooster · dog · pig

tropical year (solar year)

the exact period of time taken for Earth to complete one revolution around the sun: the interval between two successive passages of the sun through the Tropic of Capricorn, i.e., at the winter solstice in the northern hemisphere and the summer solstice in the southern hemisphere. The tropical year is equal to 31,556,925.9747 seconds, or about 365.242199 days.

leap year

a year in the **Gregorian calendar** in which an extra intercalary day has been added to account for the fraction of a day in the tropical year, making it a period of 366 days. Leap years normally occur every fourth year in the Gregorian calendar, except when the year is divisible by 100, unless it is also divisible by 400. Calendars in other cultures, e.g., the Jewish, Islamic and Chinese calendars, also have leap years which include intercalary days or months.

decade

a period of 10 years. The term decade was also used in the French Republican calendar for a period of 10 days in an attempt to decimalize the calendar, and the 10-day decade is sometimes used in meteorology and related fields.

katun

a unit of time in the Mayan calendar equal to 20 *tun*, or 7,200 days (roughly 20 years, *see* p. 87).

generation

an approximate unit of time, the interval between the birth of a parent and the birth of its child. Necessarily open to various interpretations, dependent on culture, location, and historical period, it can be anything from about 20 to 35 years.

century

a period of 100 years. Because of the absence of the year 0, the centuries are somewhat confusingly numbered: the first century C.E. comprises the years 1–100, the second 101–200, etc. So, the 20th century is generally agreed to be the years 1901–2000, making 2001 the first year of the 21st century. This did not, however, prevent worldwide celebration of the beginning of the second millennium at 00:00 on January 1, 2000.

The Chinese calendar is based on the lunar year, and works on a 60-year cycle. There are 12 months of alternately 29 and 30 days (with an intercalary month for leap years). Years are named after animals, in a 12-year cycle:

Rat
1900, 1912, 1924, 1936, 1948, 1960, 1972, 1984, 1996, 2008

Ox
1901, 1913, 1925, 1937, 1949, 1961, 1973, 1985, 1997, 2009

Tiger
1902, 1914, 1926, 1938, 1950, 1962, 1974, 1986, 1998, 2010

Rabbit
1903, 1915, 1927, 1939, 1951, 1963, 1975, 1987, 1999, 2011

Dragon
1904, 1916, 1928, 1940, 1952, 1964, 1976, 1988, 2000, 2012

Snake
1905, 1917, 1929, 1941, 1953, 1965, 1977, 1989, 2001, 2013

Horse
1906, 1918, 1930, 1942, 1954, 1966, 1978, 1990, 2002, 2014

Sheep/goat
1907, 1919, 1931, 1943, 1955, 1967, 1979, 1991, 2003, 2015

Monkey
1908, 1920, 1932, 1944, 1956, 1968, 1980, 1992, 2004, 2016

Rooster
1909, 1921, 1933, 1945, 1957, 1969, 1981, 1993, 2005, 2017

Dog
1910, 1922, 1934, 1946, 1958, 1970, 1982, 1994, 2006, 2018

Pig
1911, 1923, 1935, 1947, 1959, 1971, 1983, 1995, 2007, 2019

baktun

a unit of time in the Mayan calendar equal to 20 *katun*, or 144,000 days (roughly 394 years, *see* p. 87).

millennium

a period of 1,000 years. We are currently in the third millennium C.E., but there is controversy over when this began (*see* **century**).

man-hour

a unit of work, the amount done by one person in one hour, used in calculating costs, schedules and productivity in industry. A related term, the man-day, is the amount done by one person in one normal working day.

man-year

a unit of work, the amount done by one person in one normal working year.

equinox

the time of year when day and night are of equal length: specifically the two occasions annually when the sun apparently crosses the celestial equator. They occur around March 21 (in the northern hemisphere the vernal equinox, and the southern hemisphere the autumnal equinox) and September 23 (in the northern hemisphere the autumnal equinox, and the southern hemisphere the vernal equinox). In astronomy the term is used to define the points where the sun crosses the celestial equator, the vernal equinox as it crosses south to north, the autumnal as it crosses north to south.

solstice

one of the two occasions each year when the sun is farthest north or south of the Equator. In the northern hemisphere, these correspond to the shortest day of the year, and longest day respectively: in the southern hemisphere vice versa. The term is also used in astronomy to define the points on the ecliptic midway between the two equinoxes.

quarter day

one of the four days which divide the year into quarters for mainly financial purposes, such as the payment of rent or interest. Traditionally in the U.K. the quarter days are Lady Day (March 25), Midsummer Day (June 24), Michaelmas (September 29), and Christmas Day (December 25).

❧ Greenwich Mean Time (G.M.T.)

the standard time at longitude 0°. G.M.T. was adopted as the official standard time in the U.K. in 1880, and named after the Royal Greenwich Observatory in London. It subsequently became the basis of time zones around the world, but has been replaced by

Universal Time (U.T.). also calculated from the time at 0° longitude. A further refinement to this system of time measurement is Co-ordinated Universal Time (U.T.C.). which uses international atomic time (T.A.I.) to fix time very precisely in SI **seconds**. making adjustments to compensate for irregularities in the Earth's rotation by adding leap seconds whenever necessary.

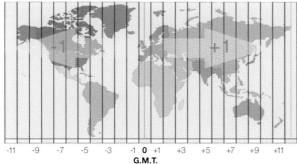

-11   -9   -7   -5   -3   -1 **0** +1   +3   +5   +7   +9   +11
G.M.T.

To minimize the number of time zones in any one country, the divisions do not follow lines of longitude exactly. Some countries, notably China, have a standard time despite spanning several time zones; others use fractions of an hour difference from U.T.

❧ time zone

one of 24 divisions of the world having a standard time in all of its area. The divisions are essentially longitudinal segments of the globe of 15° each. 12 to the east of the prime meridian (longitude 0°). and 12 to the west. with some irregularities where countries cross the divisions. Time zones to the east are ahead of Universal Time. and are designated as positively-numbered. such as Central European Time which is time zone +01: the zones to the west are behind U.T.. such as U.S. Pacific Standard Time which is time zone –08. The International Dateline follows. more or less. the line of longitude 180°. and marks where the positive and negative time zones meet. Crossing this line in an eastward direction takes you into the previous day. in a westward direction into the next day.

ephemeris time (Newtonian time; E.T.)

a former astronomical measure of time based on the unit of the ephemeris second. defined as $^1/_{31556925.9747}$ of the tropical year 1900 January 0d at 12h E.T. This was calculated from the time taken by Earth and the moon to complete a full orbit of the sun in that year. 365.24219878 days. The ephemeris second was used as the international standard in the 1960s. but has been replaced by the present SI **second**.

era

in calendrical usage. an era is a period of time in which a particular calendar has been in use. The starting date of any calendrical era is known as the emergent year. In terms of the **Gregorian calendar** in general use. we are now in the common era (C.E.).

formerly referred to as Anno Domini (A.D.). Dates previous to the emergent year of 1 C.E. are described as before common era (B.C.E.). formerly before Christ (B.C.). Calendars in other cultures have different emergent years: for example. the Jewish calendar is dated from 3761 B.C.E.. the supposed creation of the world. years being designated *anno mundi* (A.M.). or years of the world: the Islamic calendar is reckoned from the Hejirah. Mohammed's flight from Mecca to Medina. in 622 C.E.. designated in years *anno hegirae (A.H.)*.

Julian calendar

the calendar established during the reign of the Roman emperor Julius Caesar. after whom it is named. The Julian calendar was based on a 365-day year. with a leap year of 366 days every fourth year. thus making the Julian year exactly 365.25 days. This was accurate enought for it to survive from 46 B.C.E. to 1582 C.E.. when the inaccuracy of about 11 minutes each year had accumulated to an unacceptable level. and the more accurate **Gregorian calendar** was adopted.

The Jewish calendar works on a 19-year cycle, using six lengths of year: common years of 353, 354 or 355 days, and leap years of 383, 384 or 385 days. The New Year (Rosh Hashanah) falls between September 5 and October 5 in the Gregorian calendar. The Islamic calendar is based on the lunar year, and works on a 30-year cycle: leap years occur every 2nd, 5th, 10th, 13th, 16th, 18th, 21st, 24th, 26th, and 29th years.

The Jewish, Gregorian and Islamic emergent years

Jewish year (a.m.)	Gregorian year (c.e.)	Islamic year (a.h.)
5761 A.M.	2000–01 C.E.	1421 A.H.
5762 A.M.	2001–02 C.E.	1422 A.H.
5763 A.M.	2002–03 C.E.	1423 A.H.
5764 A.M.	2003–04 C.E.	1424 A.H.
5765 A.M.	2004–05 C.E.	1425 A.H.
5766 A.M.	2005–06 C.E.	1426 A.H.

❧ Gregorian calendar

the calendar adopted in 1582 C.E. to replace the **Julian calendar**. and named after Pope Gregory XIII. who decreed its use. As the year is actually about 365.242 days. the 365.25-day year of the Julian calendar progressively got out of step with the seasons. and a new system was called for. Pope Gregory decreed that 10 days be dropped from the year 1582 to make up for this anomaly. and that to prevent future inaccuracy leap years should not occur in years divisible by 100. unless they are also divisible by 400. adding up to 146.097 days every 400 years. effectively defining the Gregorian year as exactly 365.2425 days—near enough for most practical purposes.

❧ haab

the civil calendar used by the Mayans. The year was divided into 18 periods. called *uinal*. of 20 days. plus an extra five days called the *uayeb* to make a 365-day year (*see* facing page).

Components of the long count calendar

1 kin		1 day	
1 uinal	20 kin	20 days	
1 tun	18 uinal	360 days	c. 1 year
1 katun	20 tun	7,200 days	c. 20 years
1 baktun	20 katun	144,000 days	c. 394 years
1 pictun	20 baktun		c. 7,885 years

Note: Reading the long count date, the five component parts (separated by a point) are kin, uinal, tun, katun, baktun, enabling a date to be specified in an approximately 394-year period, for example, as: 13.7.18.4.1

The long count, used by the Mayans, is made up of five component parts. Longer periods, such as the calabtun, kinchiltun and alautun have also been used, but these are not included in the long count.

❧ tzolkin

the ceremonial or religious calendar used by the Mayans, based on a 260-day cycle divided into 20 periods of 13 days.

❧ long count

the calendar used by the Mayans for defining historical dates, representing the number of days from the beginning of the Mayan era (possibly September 6, 3114 B.C.E.). The basic unit of the long count is the *kin* (day), and other components of the written date are multiples of either 20 or 18 days.

ancient Egyptian calendar

a calendar used by the ancient Egyptians, based on 12 months of 30 days plus an extra five days. This 365-day year, despite attempts to introduce a leap year, continued in use until around 25 B.C.E. Uniquely, the calendar was not based on the solar year or the lunar month, but instead on the nearly coincident rising of the star Sirius and the annual flooding of the river Nile, an important event for the agriculture of the region. Because of the calendar's gradual divergence from the tropical year, however, it soon became practically useless in marking the seasons.

Months of haab
(civil calendar)
18 months of 20 days

0 pop	11 ceh		
1 uo	12 mac		
2 zip	13 kankin		
3 zotz	14 muan		
4 tzec	15 pax		
5 xul	16 kayab		
6 yaxkin	17 cumku		
7 mol	followed by		
8 chen	five extra		
9 yax	days (known		
10 zac	as uayeb)		

Days of tzolkin
(religious calendar) based on two cycles
13 numbered days and 20 named days

0 ahau	10 oc
1 imiz	11 chuen
2 ik	12 eb
3 akbal	13 ben
4 kan	14 ix
5 chicchan	15 men
6 cimi	16 cib
7 manik	17 caban
8 lamat	18 etznab
9 muluc	19 caunac

Haab and Tzolkin, the civil and religious calendars used by the Mayans.

livestock unit (LU)

the definition of livestock unit (LU) varies between different countries, and even between different areas within countries, but there is a rough general consensus that one large animal, e.g., a cow or a horse, amounts to 1 LU. The European Commission describes the LU as "a unit used to compare or aggregate numbers of animals of different species or categories." Under the European definition, a cow weighing 600 kg producing 3,000 liters of milk per year = 1 LU, a calf for slaughter = 0.45 LU, a nursing ewe = 0.18 LU, a sow = 0.5 LU and a duck 0.014 LU.

cow-calf unit

under the system of livestock units, a cow and calf are regarded as one animal until the calf is weaned, and are known together as a cow-calf unit. As such, they usually equal 1 LU, although under some definitions, they amount to 1.2 LU.

ஃ gestation period

the total time from conception to the birth of a viviparous mammal (meaning one giving birth to living offspring that have developed within the uterus). This period varies enormously according to species: the gestation period of humans, for example, is around 266 days; rats 21 days; cats 63 days and elephants 624 days.

Horse
329–346 days

Cow
273–291 days

Goat
148–156 days

Sheep
143–152 days

Pig
111–116 days

Gestation periods for farm animals. Generally, the smaller the animal, the shorter the gestation period.

incubation period

the time that an egg takes to hatch, warmed either by the heat provided by the mother bird or in an incubator. The incubation period varies according to the time of year in which the eggs are laid. In medicine, the term also means the time between exposure to infection and the first signs or symptoms of a disease.

chromosome number

a chromosome is a threadlike body found in the nucleus of a cell. Chromosomes carry nucleoprotein in the form of genes that pass on hereditary characteristics. The chromosome number is the number of chromosomes in each somatic cell of an organism, which is constant for any one species. (The reproductive cells have half this number of chromosomes.) In humans, there are 23 pairs of chromosomes in a somatic cell, each pair containing one chromosome from each parent, so the chromosome number for humans is 46.

gene density

the number of genetic loci (positions) per unit of length on the link-age map. The genetic linkage map is the graphic representation of the order of genes within a chromosome, determined by how frequently two markers (or traits) are passed on between parent and child. Recombination of genes between chromosomes can alter the pattern of inherited characteristics.

allele

an allele is any one of a number of alternative forms of the same gene occupying a given locus (position) on a chromosome. These genes govern certain characteristics of an organism, but different alleles produce different forms of these characteristics. For example, different alleles produce different-colored petals in a flower.

Morgan *see* centiMorgan.

centiMorgan

a centiMorgan (or map unit) is the distance between two genes for which the incidence of recombination is 1 percent. The American geneticist Thomas H. Morgan and members of his laboratory were the first to observe that genes are arranged on chromosomes in linear fashion, and that genes occurring on the same chromosome are inherited as a single unit for as long as the chromosome remains intact. Genes inherited in this way are said to be linked. Morgan and his colleagues discovered that there was often a breakdown of such linkage, and that recombination of genes occurred between chromosomes. The further apart the genes were, the more recombination was likely to occur.

Chemical Oxygen Demand

the amount of oxygen used in the oxidation of organic waste in water. The more oxygen is demanded by the material in the water, the bigger the C.O.D. value. C.O.D. is generally used in testing water treatment plants to gauge their efficiency in removing organic matter. It is defined as the amount of a specified chemical oxidant that reacts with a sample of waste water under controlled conditions. Waste water contains both organic and oxidizable inorganic compounds that directly or indirectly consume the available dissolved oxygen in the ecosystem, depriving wildlife of the oxygen needed to live.

biomolecule

a compound produced in living organisms, mostly comprising carbon and hydrogen, and also nitrogen, oxygen, phosphorus and sulfur, with occasionally other elements. These compounds are essential for all living things, and support the existence of different characteristics of a plant or animal. Biomolecules can be extracted for use in medicine, and some controversy has been caused by attempts to patent such use. The argument against patenting is that biomolecules are naturally occurring substances, not inventions.

biodiversity

the existence of a wide range of plant and animal life in a particular natural environment. The failure to preserve an area's biodiversity is seen by conservationists as a serious problem for the area's natural balance, leading to the extinction of some species. Loss of habitat, including deforestation, pollution and the draining of wetlands, and competition from introduced species can upset biodiversity. Farming methods also affect whether an area's biodiversity is maintained, enhanced or reduced.

Classification example	Blue whale	Explanation
Kingdom	Animalia	Whales belong to the kingdom *Animalia* because they have many cells, ingest food and are formed from a "blastula" (from a fertilized egg).
Phylum	Chordata	Animals from the phylum *Chordata* have a spinal cord and gill pouches.
Class	Mammalia	Whales and other mammals are warm-blooded, have glands to provide milk for their young, and have a four-chambered heart.
Order	Cetacea	Cetaceans are mammals that live wholly in water.
Suborder	Mysticeti	Whales in the suborder *Mysticeti* have baleen plates rather than teeth.
Family	Balaenidae	The family Balaenidae, or rorqual whales, have pleats around their throats which allow them to hold vast quantities of water (this contains their food).
Genus	Balaenoptera	The name of the genus defines a group of species that are more closely related to one another than to any other group in the family.
Species	musculus	The species is a grouping of individuals that interbreed sucessfully. The name *musculus* identifies the blue whale species.

The system of species classification devised by Linnaeus has been extended. The table shows the categories for the blue whale.

❧ species classification

a species can be defined as a group of closely related organisms that are capable of interbreeding to produce fertile offspring. A method of classification of species, or taxonomy, was developed by Carl Linnaeus in the 18th century. Each species has two Latin names, the first being the genus, or group of species, and the second is the species itself. Each species has now been further categorized by family, order, class, phylum and kingdom.

I.U.C.N. Extinction risk categories

there are eight categories on the International Union for the Conservation of Nature and Natural Resources (I.U.C.N.) "Red List" of endangered species around the globe: Extinct, Extinct in the Wild, Critically Endangered, Endangered, Vulnerable, Lower Risk, Data Deficient and Not Evaluated. A species is described as "threatened" if it falls in the Critically Endangered, Endangered or Vulnerable categories.

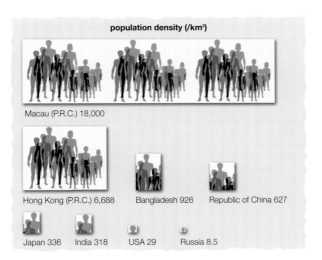

population density (/km²)

Macau (P.R.C.) 18,000

Hong Kong (P.R.C.) 6,688   Bangladesh 926   Republic of China 627

Japan 336   India 318   USA 29   Russia 8.5

❧ population density

the number of people in a given land area, usually measured as the number per square mile or kilometer. It can be a rather misleading statistic: Canada has an overall population density of 3.4 people per square kilometer, but vast areas of the north have almost no people living in them, while the province of Ontario has a population density of 11.7.

Population density statistics can be misleading. Cities have high concentrations of people and inevitably have high population density compared to countries.

population growth

population growth is usually measured annually by comparing the birth rate per 1,000 people and the death rate per 1,000 people. If in a given year there were 20 births per 1,000 people and 10 deaths per 1,000 people, population growth is 1 percent. Negative population growth occurs when deaths exceed births, zero population growth occurs when they are equal, and positive population growth occurs when births exceed deaths. Positive population growth tends to lead to exponential increases in population size, although increases also depend on the availability of resources like food, water or land.

litter

the number of offspring produced by an animal, usually a mammal, in a single birth. The figures for different groups of mammals vary considerably, and are closely related to the number of teats the animal has for suckling young. Primates tend to have one baby at a time, as do horses, elephants and whales, while dogs have four to six and pigs six to twelve.

clutch

the number of eggs laid by a single bird or laid in a single nest; or, sometimes, the brood successfully hatched from one laying of eggs.

Some of the unusual collective nouns for various types of animal
ascension of larks
bed of clams
charm of hummingbirds
crash of hippopotami
glaring of cats
harem of seals
leap of leopards
lounge of lizards
ostentation of peacocks
plump of wildfowl
rhumba of rattlesnakes
shiver of sharks
smack of jellyfish
tidings of magpies
zeal of zebras

The size of clutch varies between birds. Hens lay clutches of five or more eggs; pigeons only two. Hens can lay an average of an egg a day over the one or two years in which they are most productive. The term is also used to describe the eggs of fish and insects.

pride

a group of lions occupying a certain territory. The core of a pride consists of at least two lionesses, and sometimes as many as eighteen. They hunt together and help each other in looking after their young. Young male lions are driven out when they are about three years old. Sometimes these remain nomadic, but often they will attempt to take over a pride by displacing the existing males in a violent confrontation. Two or more males who remain together after being driven out are called a "coalition."

pack

a term used to describe a group of wild dogs or similar animals, but also the dogs used by man in hunting deer or foxes. Some wild dogs are solitary hunters, but wolves and others often hunt together, carefully coordinating their movements for the benefit of all involved.

hive

a colony of social bees, often living in a structure that is also called a hive. The colony consists of one queen bee, the only sexually developed female; several hundred drones, male bees whose sole purpose is to mate with the queen; and thousands of workers, sexually undeveloped females who feed the queen and larvae, clean the hive, guard the entrance and keep the hive cool.

swarm

a collection of bees, wasps, locusts or other social insects. In the case of bees, they are led by a queen after leaving a hive in search of another place to colonize. Swarms often cause panic among humans because of the understandable fear of large groups of stinging insects. Swarms of locusts constitute a serious danger, as they can travel as much as 120 miles (190 km) in a day and devastate crops in a matter of hours.

school

usually a group of social aquatic mammals such as dolphins, whales or porpoises. Occasionally also used for a group of hippopotami, cod, pilchards or butterfly fish, or other fish. A school of whales is known more specifically as a "pod." Toothed whales form pods more often than baleen whales. Dolphins and beluga whales can congregate in hundreds, even thousands. Schools of orca (killer whales) hunt together in a sophisticated way, much as wolves do on land.

shoal (school)

a group of fish swimming together. Fish mainly congregate to avoid predation. The risk to an individual fish is far lower in a shoal. especially a large one. Predators are often confused by large groups. and the larger the group. the fewer individual prey they manage to catch.

herd

a large group of social animals. usually grazing mammals. This term applies both to wild and to domesticated animals. including elephants. rhinoceroses. deer. cows and sheep. In the wild. the herd has a number of benefits. particularly in looking after offspring and guarding against predators. The word is also used to describe a particular breed or stock of domesticated animals.

yoke

a pair of oxen. or other draft animals like shire horses. used with a yoke to pull a plow. cart or other agricultural equipment.

flock

a group of animals. particularly sheep or birds. Sheep are kept together as flocks because they are naturally social animals. and in fact are very insecure when isolated. Their instinct for congregation is a trait inherited from the wild when being in a group offered more protection from predators. Their reaction to sheepdogs is part of this inheritance. allowing a well-trained dog to guide the sheep in a particular direction. Not all birds flock together. but those that do. like starlings or sparrows. also gain protection from predators.

❧ flight

a group of flying birds. This term particularly applies to migrating birds. such as swallows or geese. Apart from the advantage of "safety in numbers" when covering vast distances. it is also important for birds to remain in contact with other members of the species for purposes of reproduction.

Safety in numbers is what causes birds to migrate in flocks. often using recognizable routes. There are four routes—called flyways—over North America: the Atlantic, Mississippi, Central and Pacific. Songbirds, waders and birds of prey follow approximately these flyways.

93

brace

a pair of game birds. as in the phrase "a brace of pheasants." Why game birds should be grouped together like this is not clear. though one plausible reason is that just one bird was considered inadequate for a meal. and two was about right. From a purely practical point of view. it is easier to string two birds together in order to carry them home.

Physical
Sciences

chemistry

Avogadro's number (constant) (NA)

the number of atoms or molecules in 1 **mole** of a given substance. Named after Italian chemist Amedeo Avogadro (1776–1856). Avogadro's number is properly defined as the number of carbon-12 atoms in 12 grams of carbon-12. This is approximately 6.022×10^{23}. In practice, Avogadro's number can be applied to any substance to define the number of atoms or molecules needed to make up the substance's atomic or molecular mass respectively.

mole (mol)

an SI unit of quantity. One mole of a given substance contains as many elementary particles (typically atoms, molecules, ions or electrons) as there are atoms in 12 grams of carbon-12. So, 1 mole of helium atoms contains approximately 6.022×10^{23} (**Avogadro's number**) atoms, and 1 mole of oxygen molecules contains 6.022×10^{23} molecules. Since 1 molecule of O_2 contains 2 oxygen atoms, 5 moles of O_2 molecules contain 10 moles of oxygen atoms.

relative molecular weight

also called the relative molecular mass, the relative molecular weight of a compound is its molecular weight relative to $^1/_{12}$ atom of carbon-12. The relative molecular weight of a compound is calculated by adding the atomic masses (**atomic mass unit**) of each of the atoms in one molecule of that compound. For example, water (H_2O) contains two hydrogen atoms (atomic mass 1.008 u) and one oxygen atom (atomic mass 16 u). So the relative molecular weight, therefore, of water is $(1.008 \times 2) + 16 = 18.016$ u. The relative molecular weight is numerically equal to the molar mass, although the latter is measured in units of g/mol.

✌ periodic table

a table of all known elements, arranged in increasing order of **atomic number**. For each one, the periodic table traditionally displays the atomic number, the symbol of the element and its relative atomic mass. The table has existed in less sophisticated forms since 1817, but it was when Dimitri Mendeleev (1834–1907) arranged elements by atomic mass in 1869 that it began to resemble the table we know today. Interestingly, Mendeleev's table predicted the existence of certain elements that were hitherto unknown.

period

a set of elements occupying a single row in the periodic table. As one moves from left to right, the atomic number—the number of protons in the nucleus—of the element increases by one. Elements in the same period also have the same number of electron shells.

though the number of electrons in the electron shell of an element also increases by one as you move from left to right across the period. Elements horizontally adjacent to each other in the periodic table have a similar relative atomic mass, but they tend to have different properties.

periodicity

the tendency of the properties of elements to recur at intervals. Elements are divided into a periodic table precisely because it highlights their periodicity when arranged according to their atomic number (**periodic table, period, group**).

❧ group

The periodic table has undergone a number of transformations ever since it was first suggested. This is the version currently in use.

a set of elements occupying a single column of the **periodic table**. There are 18 groups, or columns, in the periodic table, and elements that fall within the same group tend to have similar properties (e.g., whether they occur as solids, liquids or gases at a given temperature). For example, group 18 is populated by the noble gases helium, neon, argon, krypton, xenon and radon, which all show the similar property of being relatively unreactive. Elements within the same group have the same number of electrons in their outer electron shell.

1 H																	2 He
3 Li	4 Be											5 B	6 C	7 N	8 O	9 F	10 Ne
11 Na	12 Mg											13 Al	14 Si	15 P	16 S	17 Cl	18 Ar
19 K	20 Ca	21 Sc	22 Ti	23 V	24 Cr	25 Mn	26 Fe	27 Co	28 Ni	29 Cu	30 Zn	31 Ga	32 Ge	33 As	34 Se	35 Br	36 Kr
37 Rb	38 Sr	39 Y	40 Zr	41 Nb	42 Mo	43 Tc	44 Ru	45 Rh	46 Pd	47 Ag	48 Cd	49 In	50 Sn	51 Sb	52 Te	53 I	54 Xe
55 Cs	56* Ba	71 Lu	72 Hf	73 Ta	74 W	75 Re	76 Os	77 Ir	78 Pt	79 Au	80 Hg	81 Tl	82 Pb	83 Bi	84 Po	85 At	86 Rn
87 Fr	88+ Ra	103 Lr0	104 Rf	105 Db	106 Sg	107 Bh	108 Hs	109 Mt	110 Ds								

57 La	58 Ce	59 Pr	60 Nd	61 Pm	62 Sm	63 Eu	64 Gd	65 Tb	66 Dy	67 Ho	68 Er	69 Tm	70 Yb
89 Ac	90 Th	91 Pa	92 U	93 Np	94 Pu	95 Am	96 Cm	97 Bk	98 Cf	99 Es	100 Fm	101 Md	102 No

* Lanthanides Series
+ Actinide Series

❧ isotope

atoms of a particular element that have the same number of protons, but a different number of neutrons. Carbon-12, for example, has 6 protons and 6 neutrons; carbon-14, however, has 6 protons and 8 neutrons. The word isotope derives from the Greek word meaning "the same place" and is used because isotopes of a particular element occupy the same place in the **periodic table**. The difference in ratio between the protons and neutrons of the isotopes of a particular element means that isotopes of given elements often display different degrees of nuclear stability and some—known as radioactive isotopes—are subject to nuclear decay.

allotrope

any of the different forms in which an element may occur. Allotropes of oxygen, for example, include O_2, where two oxygen atoms are bound together, and O_3 (ozone) where three oxygen atoms are bound together. Similarly, allotropes of carbon include diamond, a three-dimensional crystal where each carbon atom is bound to four neighbors, and graphite, a layered structure where each atom has strong bonds to three neighbors within the same layer. Allotropes are not the same as the different states of matter—solid, liquid and gas—and an allotrope may be able to exist in one, two or all three states, depending on other factors.

valency

a measure of the power of an element to combine with other elements to form a compound, specifically its power to combine with an atom of hydrogen. An element's valency is equal to the number of spaces left in the electron shells of its atom. So, the valency of oxygen is 2, because it requires 2 electrons to fill up its outer electron shell. As the hydrogen atom has 1 electron, it therefore requires 2 hydrogen atoms to react with 1 oxygen atom to form H_2O, or water.

bond energy

the amount of energy released when a chemical bond is formed, equivalent to the amount of energy required to break that bond. Since the energy required to make or break chemical compounds is generally converted into heat, bond energy tends to be measured in kJoules—the amount of energy required to heat 1 liter of water by 1°C. The bond energy of the hydrogen–oxygen bond, for example, is 110 kJoules.

dipole moment

a measure of the extent of electrical separation of a molecule. Although the overall charge of a molecule may be neutral, it does not necessarily follow that its electrical charge is equally distributed, and this differential can cause magnets or electrical currents to exert a torque on the molecule. For a diatomic molecule, the dipole movement is equal to the positive charge multiplied by the distance between two charges; more complex molecules may have a dipole moment (e.g., water) or not (e.g., methane), depending on the arrangement of the chemical bonds. The dipole movement can affect the **rate of diffusion** of a particle.

electronegativity

a measure of the tendency of an atom to attract electrons and so form a covalent bond. Several scales are used to measure electronegativity: the Pauling scale, the Mulliken scale and the Alfred-Rochow scale. Elements toward the top right-hand corner of the periodic table tend to have a higher degree of electronegativity than those at the bottom left-hand corner, and electronegativity decreases as you go down the periodic table. The most electronegative element is fluorine; the least electronegative is francium.

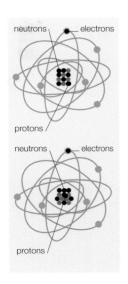

The isotope of carbon-12 (top) has 6 protons and 6 neutrons. The isotope of carbon-14 (bottom) has the same number of protons, indicating that it is still carbon, but 8 neutrons.

🍃 pH

a measure of the acidity or alkalinity of a solution. according to its concentration of hydrogen ions. pH stands for potentiality of hydrogen. and is equal to $-\log_{10}C$ (**logarithm**) where C is the concentration of hydrogen in moles per liter. A pH of 7 indicates that a solution is neutral. A higher pH indicates a greater degree of alkalinity: a lower pH indicates a greater degree of acidity.

equivalent weight

the atomic weight of an element divided by its **valency**. The unit of equivalent weight is the **atomic mass unit** (amu). The equivalent weight of an element indicates how much of that element is required to react with another. For example. the equivalent weight of hydrogen is 1.008 amu: the equivalent weight of oxygen is 7.9997 amu: water (H_2O) is therefore about 11.2 percent hydrogen by weight.

octane number

the percentage of octane in an octane–heptane mixture. Isooctane. for example. has an octane rating of 100. The octane number of a fuel is a measure of how smoothly that fuel burns. or in the case of motor fuel. a measure of its anti-knock performance. and indicates that the fuel burns in the same way as an octane–heptane mixture of that particular octane number.

katal (kat)

the SI unit of catalytic activity (**catalysis**). A catalyst is defined as having an activity of 1 katal if it allows a reaction to proceed at a rate of 1 mole per second. Katal is pronounced "cattle."

catalysis

the process of speeding up the rate of a chemical reaction by means of a catalyst. A catalyst is a substance that increases the rate of a chemical reaction. without undergoing any chemical change itself: it does not. as is commonly misunderstood. make the reaction happen. The unit of catalysis is the **katal**.

reactivity

the rate at which an element or other substance tends to undergo a chemical reaction. The reactivity of a substance is essentially determined by its atomic or molecular structure. but it can be increased or decreased by changing its physical characteristics— e.g.. grinding a substance to a powder may increase its reactivity.

🍃 reversible

a reaction in which the products may react back to form the original reactants. When a reversible reaction occurs within a confined space. an **equilibrium** is reached where the quantity of reactants and products does not change. If a quantity of one or other of these

Litmus paper will indicate the pH of a particular solution by turning a certain color when it comes in contact with that solution.

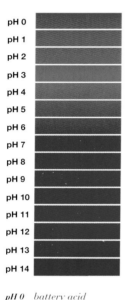

pH 0
pH 1
pH 2
pH 3
pH 4
pH 5
pH 6
pH 7
pH 8
pH 9
pH 10
pH 11
pH 12
pH 13
pH 14

pH 0   *battery acid*
pH 1   *sulfuric acid*
pH 2   *lemon juice, vinegar*
pH 3   *orange juice, soda*
pH 4   *acid rain (4.2–4.4)*
      acidic lake (4.5)
pH 5   *bananas (5.0–5.3)*
      clean rain (5.6)
pH 6   *healthy lake (6.5)*
      milk (6.5–6.8)
pH 7   *pure water*
pH 8   *sea water, eggs*
pH 9   *baking soda*
pH 10   *Milk of Magnesia*
pH 11   *ammonia*
pH 12   *soapy water*
pH 13   *bleach*
pH 14   *liquid drain cleaner*

is then removed, the pro-
portions will continue to
change until an equilib-
rium is again reached.

$$2\,NO_2\,(g) \rightleftharpoons N_2O_4\,(g)$$

degrees of freedom

any of the ways in which an individual particle may move. Each
degree of freedom of a particle will contain on average the same
amount of thermal energy, equal to the temperature multiplied by
the fundamental Boltzmann constant. Heisenberg's uncertainty
principle states that the amount of energy in any degree of freedom is never equal to zero.

*A chemical reaction
describing the reversible
reaction between nitrogen
dioxide and dinitrogen
tetroxide. The (g) stands for
gas; many reactions are
reversible in some states but
not others.*

equilibrium

a state during which a process such as a chemical reaction and its
opposite are occurring at equal rates, so that there is no net change
in the quantities of the reactants and the products. For example, in
the Haber-Bosch process, hydrogen and nitrogen react to form
ammonia, which in turn reverts back to the constituent parts.
Equilibrium is reached when the rate of production of ammonia
equals the rate of its decomposition.

buffer solution

a solution which resists change to its **pH** when a small amount of
acid or alkali is added to it. Often consisting of a weak acid and its
salt, buffer solutions maintain pH as a result of the acid and the
salt reacting and entering a state of **equilibrium**. A solution of
carbonic acid and bicarbonate exists as a buffer solution in blood
plasma to maintain a pH of about 7.4.

molarity (M)

a measure of concentration. Molarity is the proportion of a substance in a solution, and is defined in terms of **moles** per liter. So,
a solution of 1M has 1 mole of solute per liter of solvent; a solution of 2M has 2 moles of solute per liter of solvent; and a solution
of 1mM has 1mmol of solute per liter of solvent.

molality (m)

a measure of concentration. Molality, like molarity, is the proportion
of a substance in a solution, but defined in terms of moles per kilogram. A solution of 1m has 1 mole of solute per kilogram of solvent;
a solution of 2m has 2 moles of solute per kilogram of solvent; and
a solution of 1mm has 1 mmol of solute per kilogram of solvent.

osmotic pressure

a measure of the force per unit of area required to stop the process
of osmosis. If the pressure exerted on the area of high concentration
is increased relative to that exerted on the area of low concentration, the osmotic process will slow down and eventually stop.

water molecules   solute molecules

semi-permeable
membrane

only water molecules
can pass through the
membrane, so the
osmotic pressure
increases

*Osmosis is a process
whereby the solvent from a
solution with a relatively
high concentration of solute
passes through a semi-
permeable membrane to a
solution with a relatively
low concentration of solute.
Such a membrane is
permeable to the solvent,
but not to the solute. The
process continues until the
solution is of the same
concentration on both sides
of the membrane.*

*This phase diagram of water
shows how temperature and
pressure affect the state of
matter of a substance. Using
the phase diagram for a given
substance, therefore, you can
predict what state of matter
the substance will be at if you
know its temperature at a
given pressure. Water is most
unusual in that it melts under
(immense) pressure.*

❧ rate of diffusion

the rate at which the natural
and spontaneous spreading of
particles occurs, e.g., smoke
spreading in air or a drop
of colored liquid in water.
Osmosis is an important form
of diffusion for humans as it is
the process by which water
enters the cells of the body.

diffusion gradient

any measure of diffusion in a given direction. If there is no
gradient, there is no apparent change in concentration (as is the
case with thermal diffusion at a constant temperature).

diffusivity

a measure of a substance's tendency to be diffused. Substances
with high diffusivity will tend to have a high rate of diffusion
under favorable circumstances; substances with low diffusivity
will tend to have a low rate of diffusion.

lattice types

any of the various formations which the atoms of a crystal may
take. There are 14 different lattice types, which include the face-
centered cubic lattice (fcc), the body-centered cubic lattice (bcc)
and the sodium chloride lattice (NaCl).

unit cell

the smallest group of atoms in a crystal which display the overall
symmetry of that crystal. The unit cell is repeated in three dimensions to form the lattice of a crystal.

phonon

a quantum measure of vibrational energy. The atoms in a crystal
are in a constant state of vibration, and the quantums of energy
that the atoms emit behave in some respects like particles.
Interactions between phonons and electrons are thought to be
responsible for phenomena such as **superconductivity**.

❧ phase diagram

a graph which displays the effect of both temperature and pressure on the three states of
matter of a substance, or the effects of temperature and composition on a mixture of two
(occasionally three) different substances. The
first type is divided into three sections, one for
each state of matter—solid, liquid and gas—
or more (if the substance has **allotropes**).

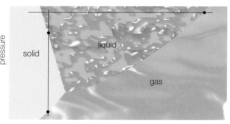

pressure

solid

liquid

gas

temperature

electricity and magnetism

resistance

the ability of a material to resist an electric **current** passing through it. When the material is a conductor, resistance is the ratio of the **potential difference** between the ends of the conductor to the current flowing through it. Resistance therefore equals the potential difference in **volts** divided by the current in **amperes**.

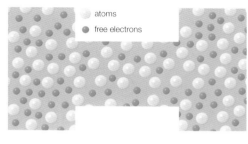

atoms
free electrons

Resistance depends on the ability of electrons to more within a conductor. The greater the diameter of a wire, the greater the electron (current) flow.

superconductor

the term for a material with a very high level of conductivity at a very low temperature. The specific resistance of metals decreases with temperature, and at a given temperature, known as the "transition temperature," approaching absolute zero ($-273°C$), the electrical resistance of certain substances becomes zero. Hence, a current flowing through such a metal at this temperature will continue almost indefinitely. Superconductors include the metals tin, aluminum and lead; a number of alloys; some semiconductors that have been heavily doped (for silicon, usually phosphorus or arsenic is used), and some ceramic compounds containing copper and oxygen. The latter, known as cuprates, are also called high-temperature superconductors, even though the reference to high temperature is only relative to the other types of super-conductor—superconductivity still occurs at a temperature far below room temperature.

ohm

the unit of electrical resistance, named after German physicist Georg Simon Ohm (1789–1854), who formulated Ohm's Law: the **potential difference** (in **volts**) across a conductor is equal to the **current** through it (in **amperes**) multiplied by the **resistance** (in ohms). The ohm is the oldest electrical unit used, dating from the mid-19th century.

impedance

the ratio between the **root mean square** values (effective values) of **voltage** and **current** in a device or circuit when an **alternating current** (A.C.) is applied, measured in **ohms**. Impedance is a complex concept involving **resistance**, conductance and inductive and capacitive reactance. The current will normally be out of phase with the voltage. The term impedance is sometimes used as a synonym for resistance in a D.C. circuit.

resistivity

also called "specific resistance." the **resistance** of a unit cube of a substance. This is usually a measurement in ohms per cubic meter (the ohm meter). and it is constant for a given substance at a given temperature. It is used to compare the abilities of different substances to resist a current.

current

the passage of electricity through a conductor. In a solid. usually a metal. the carriers of electricity are electrons. Electrons actually flow from negative to positive, but for historical reasons conventional circuit diagrams show them flowing from positive to negative. Current is measured in amperes. It is related to electrical power and **potential difference** and 1 watt is the energy expended per second by a steady current of 1 ampere flowing in a circuit across which there is a potential difference of 1 volt.

direct current (D.C.)

an electric current that flows in one direction only. Normally only applies to low-voltage battery-operated devices. e.g.. a flashlight. portable C.D. player. or cell phone.

ampere ("amp")

the SI base unit of **current**. usually shortened to "amp." named after French physicist André Marie Ampère (1775–1836). It was defined by the Conférence Générale des Poids et Mesures (which defines SI units) in 1948 as "that constant current which. if maintained in two straight parallel conductors of infinite length. of negligible circular cross-section. and placed one meter apart in a vacuum. would produce between these conductors a force equal to 2×10^{-7} newtons per meter of length."

The electricity used in our homes has a sinusoidal waveform, but other waveforms are possible: square and sawtooth waves are produced by electronic oscillators, for example.

amp hour (ampere hour; Ah)

the amount of energy charge in a battery that will allow 1 ampere of **current** to flow for 1 hour. The milliampere hour (mAh) is often used when describing batteries for portable computers. providing an indication of how long a battery will last before it needs recharging.

alternating current (A.C.)

a current that periodically reverses its direction. generally in a regular fashion. This means that the flow of electrons is constantly reversing. Many electrical devices would not work without A.C. and it is the mains power supply of most countries. In most cases the frequency of the cycle is 60 times a second. or 60 hertz. but sometimes it is 50 Hz. A.C. circuits can be irregular. particularly in the case of analog signals (the human voice or music) interpreted by audio amplifiers.

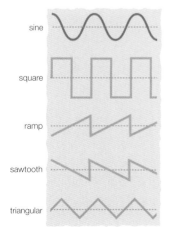

sine

square

ramp

sawtooth

triangular

root mean square (RMS)

the sum of the square roots of a series of values, divided by the number of values. In electricity, this has particular relevance to A.C. circuits: the RMS value, also known as the effective value, is the mean power level of an A.C. circuit. It is calculated from instantaneous values over a complete cycle. In the U.S. the RMS value is 110 volts, in Britain it is 240 volts.

hertz (Hz)

the SI unit of frequency, equal to one cycle per second. The frequency of a cycle is particularly relevant in the case of an **alternating current**, and the electromagnetic waves produced by it. Electromagnetic radiation is the basis of wireless communication, and includes radio and infrared waves. The wavelength is inversely proportionate to the frequency. So the higher the wavelength, the lower the number of hertz. Computer processor speeds are measured in megahertz or gigahertz. The unit is named after German physicist Heinrich Hertz (1857–94).

potential difference

the voltage in a circuit. If work is done to pass a charge between two points, then a potential difference exists between them, and this is measured in **volts**.

volt

the unit of **potential difference** or electromotive force. One volt is the potential difference between two points in a circuit when 1 **joule** of energy is expended in making 1 **coulomb** of electricity pass from one point to the other. One volt can also be expressed as using 1 **watt** of power at a constant current of 1 **ampere**, or as the potential difference across a resistance of 1 ohm when a current of 1 ampere is passed through it. The unit is named after Italian physicist Alessandro Volta (1745–1827).

electron-volt (eV)

the increase in energy of an electron falling through a **potential difference** of 1 **volt**. One electron-volt = 160.206×10^{-21} joules. It is an extremely small unit of energy, and a more convenient unit is the mega-electron volt, giga-electron volt or even the tera-electron volt. Energy in atomic and nuclear physics is commonly expressed in electron volts. Nuclear fission releases about 200 MeV of energy. The world's highest-energy particle accelerator at Fermilab in the U.S. accelerates protons to nearly 1 TeV (1 trillion electron volts; 10^{12} eV).

Faraday's constant

the amount of electric **charge** required to deposit 1 **mole** (**Avogadro's number** of molecules) of any mono-valent element.

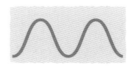

long wavelength, low frequency, low energy

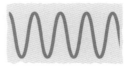

short wavelength, high frequency, high energy

The lower of the two waves has a shorter wavelength but a higher frequency measured in hertz, than the wave at the top.

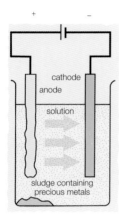

+   –

cathode
anode

solution

sludge containing
precious metals

Electrolysis is used for electroplating objects, some types of purification and performing certain chemical reactions. The anode accepts electrons (either from atoms attached to it, which then become positive ions, or from negative ions in the solution, which become neutral atoms) while the cathode neutralizes positive ions in the solution by donating electrons to them. In electroplating, it will be metal ions that form a very thin coat of solid metal on the surface of the cathode.

equal to about 96,485 coulombs. The unit is also known simply as Faraday (symbol F), and is named after British electrochemist and physicist Michael Faraday (1791–1867), who first identified it. Faraday's constant is particularly important in electrolysis. His first law of electrolysis states: "The amount of a substance deposited on each electrode of an electrolytic cell is directly proportional to the current through the cell."

charge

the existence of an excess or deficiency of electrons in a body. A body with an excess of electrons is said to be positively charged, and one with a deficit negatively charged. Charge may be quantified in terms of coulombs.

coulomb

the SI unit of electrical or electrostatic charge: the amount of electricity transferred when a current of 1 **ampere** flows for 1 second. Named after French physicist Charles Coulomb (1736–1806), who formulated Coulomb's Law. This law states that the mutual force between two point electrostatic charges is proportional to the product of the charges, and inversely proportional to the square of the distance between them. The law also applies to the force between magnetic poles. One coulomb = 6.3 $\times 10^{18}$ elementary charges (an elementary charge is the charge on one electron or proton).

capacitance

the ability to store an electrical charge. A capacitor is an electronic component usually with two plates separated by a small distance. When a voltage is passed across the capacitor, a positive charge is stored on one plate, and an equal negative charge on the other. The capacitance, measured in **farads**, is the ratio between the charge (measured in coulombs) and the voltage.

farad

the SI unit of **capacitance**, equivalent to a charge of 1 coulomb divided by 1 volt. In practice, because it is such a large unit, the more convenient microfarad (10^{-6} farad), nanofarad (10^{-9} farad) or picofarad (10^{-12} farad) are used.

inductance

the property of an electric circuit opposing any change in electric current. It is measured in henrys. A change in a circuit causes an electromotive force (E.M.F.) opposing the change. As an A.C. circuit is constantly reversing, changes in its magnetic field induce counter E.M.F. Self-inductance occurs when the voltage is induced in the conductor that carries the current, and mutual inductance when the voltage is induced in another conductor nearby.

henry

the SI unit of **inductance**. If the current in a circuit changes at the rate of 1 ampere per second and the consequent electromotive force is 1 volt, the inductance of the circuit is 1 henry. Henrys are used to describe both self-inductance and mutual inductance.

dynamo

a mechanical device that produces electrical energy, originally devised by **Faraday**. A rotating armature or coil fixed around or inside a magnet induces an electromotive force (E.M.F.); the continuous rotation of the armature in a uniform magnetic field produces an alternating E.M.F. and current can be supplied. To convert an **alternating current** to **direct current**, a commutator, which reverses the current in the negative part of the cycle, is used.

eddy current

induced current created in the core of an electromagnet or transformer by changing magnetic fields. The heat produced is an indication of inefficiency, or wasted energy, and measures can be taken to reduce it. One way is to laminate the metal core, with insulators placed between each thin sheet of metal.

rheostat

a variable resistor. Usually made from a coil of wire with a contact that slides against it. The current arrives at one end of the coil and leaves from the point where the contact is touching the wire. The longer the distance, the greater the resistance. Common uses of rheostats (also called potentiometers) are in dimmer switches for lights or volume controls on audio equipment, though such analog devices are now often replaced with digital alternatives.

turn

a loop of wire in a coil (e.g., of a **rheostat** or **dynamo**). Turn can also refer to the rotary movement of a potentiometer; some potentiometers are described as "multi-turn" because they have a helical structure and require several 360-degree turns to take them through the whole resistance range. An ampere-turn is the **current** flowing in a coil multiplied by the number of turns.

gauss

a unit of magnetic induction or magnetic flux density (magnetic field) in the C.G.S. system. One gauss is 1 **maxwell** per square centimeter. A magnetic field of 1 oersted produces an induction of 1 gauss in air, and of μ gauss in a medium whose magnetic permeability is μ. One gauss is equivalent to 10^{-4} **tesla**.

magnetic bottle

the use of magnetic fields to "bottle up" a plasma during controlled thermonuclear reactions. It would be impossible to hold a plasma

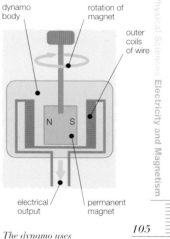

dynamo body

rotation of magnet

outer coils of wire

N   S

electrical output

permanent magnet

The dynamo uses magnetism to provide electromotive force, and hence current.

in any normal solid container. because it is too hot. The plasma has a magnetic charge and the "bottle" has an opposite charge, thereby holding it in place.

magnetic flux density

the magnetic flux per unit area, also known as magnetic induction (though this latter term is normally used to mean a process in which a material is "magnetized," usually by placing it within a magnetic field). The magnetic flux is the size of the magnetic field, and is measured in **webers** (joules per ampere). Magnetic flux density is measured in **teslas** (webers per square meter). The magnetic field strength, H, is altered by a substance placed in it, and acquires a new value B (the flux density) because of the magnetic properties of the substance. If the ability of the substance to become magnetized, the "magnetic permeability," is μ, then $B = \mu H$.

maxwell

a unit of magnetic flux in the C.G.S. system. It is the flux per square centimeter, perpendicular to a magnetic field with intensity of 1 **gauss**. The unit was named after British physicist James Clerk Maxwell (1831–79), who linked a set of equations describing the laws of electricity and magnetism, demonstrating the mathematical relationship between electric and magnetic fields. He recognized that light was a form of electromagnetic radiation, a theory that was later confirmed by Hertz.

tesla

the SI unit of **magnetic flux density**, amounting to one weber per square meter. Named after Serbian-American physicist Nikola Tesla (1846–1943).

magnetic field strength (H)

also known as "magnetic intensity" and "magnetizing force," and measured in **amperes** per meter. It is defined as a **vector** quantity whose magnitude is the strength of a magnetic field at a point in the direction of the magnetic field at that point. It is proportional to the length of a conductor and the amount of electrical **current** passing through the conductor.

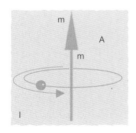

Magnetic moment is described mathematically as m = IA, where I is the current and A is the area enclosed by the current loop.

❧ magnetic moment

for a magnetic field of unit magnetic field strength, the torque that is required to hold a magnet perpendicular to the field. It is also expressed as the product of the pole strength of a magnet and the magnetic length (the distance between the poles). Electrons and thus atoms have magnetic moments, some of which are very small, but others, like those of iron, are very large. The size of an electron's magnetic moment is expressed in Bohr magnetons, named after Danish physicist, Niels Bohr (1885–1962).

temperature

absolute zero

the temperature at which an object will have no heat energy at all, and its constituent atoms therefore reach what is called their "ground state" (they can never actually stop moving completely). Since it is impossible to have negative energy, absolute zero is (in theory) the lowest possible temperature. In practice, the laws of thermodynamics make it impossible ever to reach absolute zero, but scientists have managed to come within a billionth of a **kelvin**. The temperature above absolute zero is sometimes called the absolute temperature.

helium scale

a scientific technique for measuring very low temperatures (below 5 K) using the vapor pressure of helium.

ᴥ Celsius (°C)

a temperature scale, with 0°C fixed at the freezing point of pure water, and 100°C at its boiling point (both at standard atmospheric pressure). It was originally called centigrade (or occasionally centismal), but was officially renamed in 1948 to honor the Swedish astronomer Anders Celsius, who invented it in 1742.

ᴥ kelvin (K)

the SI scale of temperature. One kelvin is equal to 1° Celsius but 0 K is at absolute zero, equal to −273.16°C (so water freezes at 273.15 K). The full name of the unit (named after British scientist William Thomson, later Lord Kelvin) should be written with a lower-case k, but the abbreviation is a capital K and neither should ever be used with "degrees" or the ° symbol.

The Fahrenheit, Celsius, kelvin and Rankine temperature scales.

ᴥ Fahrenheit (°F)

a temperature scale proposed in 1724 by the German physicist Gabriel Fahrenheit. His initial scale was calibrated from the melting point of a mixture of equal weights of salt and water at 0°F, and the temperature of horse blood (assumed to be the same as a human's) at 96°F, and the melting and boiling points of water at 32°F and 212°F, respectively. Unfortunately, his measurements were rather inaccurate, and the scale was recalibrated after his death using the freezing and boiling points of pure water as the fixed points.

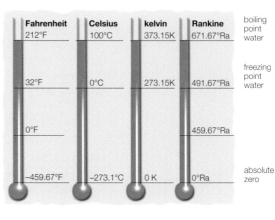

❧ Rankine (°Ra)

a rarely used scale with the same degree size as Fahrenheit but (like the kelvin scale) with 0°Ra at **absolute zero**. Named after the Scottish physicist William Rankine, who proposed it in 1859. 0°F is 459.67°Ra. and the freezing point of water is 491.67°Ra.

Reaumur (°R)

a temperature scale (now increasingly rarely used) with 0°R at the freezing point of water (as Celsius). but 80°R at the boiling point. so that 1°R is 1.25°C. Originally proposed by the French scientist of the same name in 1731. the size of the degree is a function of the thermometer he originally measured it with.

infrared

the part of the electromagnetic spectrum between microwaves and visible light (between about 300 GHz and 400 THz). It is emitted by any object with a temperature above **absolute zero**. and is felt by a human as heat.

boiling point

the temperature at which a substance changes from liquid to gas (also called the condensation point. since the change from gas to liquid occurs at the same temperature). It varies considerably with pressure. being highest at low pressure.

freezing point

the temperature at which a substance changes from liquid to solid (or vice versa. so that it is sometimes called the melting point— again. as with **boiling point**. the temperature is the same). Most substances freeze more easily (that is. at a higher temperature) when under high pressure. Water. however. behaves rather unusually: it expands when cooled below 4°C. so that ice melts under pressure. The phenomenon is often said to explain how skates work. though in fact the pressure of the skater on the ice makes a difference to the melting point of only a fraction of a degree. Ice above a temperature of about −150°C always. in effect. has a very thin layer of water on the surface. which explains the low friction. Measurements suggest that ice is at its most slippery at around −5°C to −10°C.

supercooling

the name for cooling a liquid past its melting point without letting it solidify. This is possible because solidification occurs most easily around a "nucleus"—an existing solid particle (which can include the container the liquid is inside). In the absence of a suitable nucleus. supercooling can occur. although there is a fixed point (−39°C. for water) at which each liquid freezes without requiring a nucleus in a process called "dynamic arrest."

sublimation point

a few substances (such as the element iodine) pass directly from solid to gas on heating (and vice versa on cooling), and cannot normally exist in a liquid state. This process is called sublimation, and the temperature at which it occurs is the sublimation point.

triple point

the combination of temperature and pressure at which some substances can exist simultaneously in gaseous, liquid and solid forms, all in equilibrium. (The triple point of water is at $0.01°C$ and 612 Pa, with no other gas present.)

Standard Temperature and Pressure (S.T.P.)

a standard to allow replication of scientific experiments in controlled, identical conditions. It is taken to be $0°C$ and 101.325 Pa. A similar concept, R.T.P. (Room Temperature and Pressure), may be used to mean a temperature of about $20°C$ and ambient atmospheric pressure—far easier to achieve without special equipment.

specific heat capacity (c)

the amount of energy needed to raise the temperature of 1 kg of a substance by 1 K, without changing its state. This condition is important, as very large amounts of energy can be required to alter the state of a material (heating 1 kg of water by 1 K uses 4.2 kJ of energy; converting 1 kg of ice at $0°C$ to water at $0°C$ requires 334 kJ). The amount of energy needed to change the state of 1 kg of a substance, without altering its temperature, is called the latent heat. Energy must be added to change from liquid to gas or solid to liquid, while the opposite changes require energy to be removed.

temperature gradient

the rate at which the temperature changes with distance across a material (including air). Easily calculated for a single substance by dividing the total temperature difference by the distance from one side to the other, though this assumes that no heat is lost from the sides and that the heat travels across in a straight line. For a multi-layer object (e.g., double-glazed windows), each layer can (and usually does) have a different temperature gradient.

thermal conductivity

a measure of a substance's ability to allow heat to pass through. Calculated by multiplying the flow rate of heat energy by the thickness of material, and then dividing this product by the cross-sectional area of material multiplied by the temperature difference. For a **temperature gradient** of 1, this becomes simply heat flow divided by cross-section, with the SI unit being watts per meter-kelvin. The thermal resistivity of a substance is simply 1 divided by the thermal conductivity.

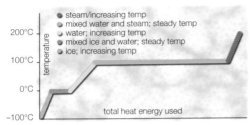

- steam/increasing temp
- mixed water and steam; steady temp
- water; increasing temp
- mixed ice and water; steady temp
- ice; increasing temp

200°C

100°C

0°C

−100°C

temperature

total heat energy used

Total heat energy graph for converting ice at −100°C to steam at 200°C. About two-thirds of the total energy is used in the phase change converting water at 100°C to steam at 100°C.

thermal diffusivity

the ratio of a substance's thermal conductivity to its specific heat capacity. Thermal diffusivity is a measure of how quickly an object will match the temperature of its surroundings. The higher the thermal diffusivity, the faster it absorbs heat compared to the amount it needs, so the faster it will adjust to new surroundings.

frigorie

a unit of refrigeration, defined as the extraction of heat at a rate of 1 kilogram calorie per hour. The name is modeled on "calorie," using the Latin *frigus* (cold) instead of *calor* (heat), but the unit is too small for most practical purposes.

ton

the amount of heat that must be extracted to completely freeze a ton of water at 0°C over a 24-hour period. This works out as approximately 200 **Btu** per minute, and a ton is now defined as precisely that value, or around 3.5 kW.

refrigerant tonne

the metric equivalent of the above, defined as one kilogram calorie per second (3,600 **frigorie**, or just under 3.9 kW).

tog

a unit of thermal resistance, defined as the temperature difference in degrees Celsius (between the warm and cold sides) divided by 10 times the flow of heat energy (in watts per square meter) from one side to the other. Used mainly for bedding and winter clothing, typical tog values range from about 5 (for lightweight summer bedding) up to 15 (a very heavy winter duvet).

Clo

another unit of thermal resistance, also used primarily for textiles. 1 clo is a little over 1.5 tog. One clo is the amount of insulation that maintains a heat-flow of 50 kilogram calories per hour between 70°F on one side and body temperature on the other.

U-factor (U-value)

a unit of thermal conductivity for building materials, measured as the heat loss in **Btu** per square foot per hour for a temperature gradient of 1°F per unit thickness. The lower the U-value, the better the material works as an insulator. The equivalent unit of thermal resistivity is the R-value, equal to 1 divided by the U-value.

R.S.I. value

the SI unit of thermal insulation (so named from the previous R-value), measured in kelvin meters-squared per watt ($K.m^2.W^{-1}$). The R-value of a substance is about 5.7 times the R.S.I. value (so an R-value of 15 is roughly equal to an R.S.I. value of 2.6). Since this is the inverse of the U-value, the better the insulation provided, the higher the number.

Btu (British thermal unit)

technically a unit of energy, but used primarily in the context of heat (including central heating systems and steam turbines). Defined as the amount of energy required to heat 1 lb of water by 1°F, although the starting temperature of the water being heated changes the value of the Btu slightly (the higher the temperature, the lower the energy in 1 Btu). One Btu is around 1,055 joules, so one Btu per hour is a little under 0.3 watts. A therm is equal to 100,000 Btu, although of course you must still specify the temperature in order to know the exact value.

Ctu (Celsius thermal unit)

a rather strange attempt to "update" the **Btu** by partial metrication, equal to the amount of energy necessary to heat 1 lb of water by 1°C (= 1.8 Btu).

degree-day

a measure of cumulative warmth, relative to a specified base temperature, calculated by multiplying the increase in temperature by the number of days for which it was maintained. For example, if a base of 15°C is used, 10 degree-days would be a single day at 25°C, 5 days at 17°C or 10 days at 16°C. The unit is used to measure the heating requirements of a building over a winter, or the additional warmth provided by a greenhouse, and for similar purposes.

thermometer

a device for measuring temperature. The most common type is a sealed glass tube with regular markings, containing a liquid which expands steadily on heating (usually mercury or dyed alcohol). Other types include thermistors (electronic devices that vary their resistance with temperature), **thermocouples** and the ornamental Galileo thermometers, which rely on the density of water varying with temperature. A thermograph is a thermometer that keeps a constant record of temperature—for example, by drawing a graph or using a computer.

thermostat

a feedback system used to control temperature (for example, in fridges or boilers). It consists of a thermometer (often a thermocouple) and an automatic control system (either electronic or mechanical) for the source (or extractor, in a fridge) of heat.

thermocouple

a cheap, simple and reliable form of thermometer, a thermocouple consists of two lengths of different metals connected at one end,

Night temperature	Tog
15°C to 8°C	3–5
10°C to 0°C	5–8
3°C to –10°C	7–10

NOTE: *Sleeping bags used by backpackers outdoors have their tog values equated to a comfort temperature rating.*

with an instrument for measuring voltage joined to the other ends. Changing the temperature at the joint causes the voltage across it to change in proportion, and the temperature can therefore be calculated to within a couple of degrees.

pyrometer

a thermometer designed for use at very high temperatures, e.g., in pottery kilns, industrial processes such as the blast furnace, or within a volcano.

entropy

a measure of the amount of energy (in a collection of objects) that is unavailable to do work. In thermodynamics, entropy is defined as the difference between the amount of energy in the system (collection of objects) and the amount of energy available to do work divided by the absolute temperature. It is not necessarily the same as heat energy, since some heat energy can be transformed into other forms of energy (and thus made to do work), provided there is a temperature gradient. It is never possible to convert all of a system's heat energy to work, however (which is why perpetual motion machines are impossible). The laws of physics suggest that total entropy can only increase—your fridge can reduce the entropy of its contents, but only by increasing the entropy of its surroundings by a greater amount. This phenomenon leads to the famous "heat death" of the universe, when the entire universe will reach a state where everything is the same temperature and no energy is available to do anything, although there is some doubt over whether the theory can really be applied to the universe in this way.

With four coins, the most likely result is two heads and two tails (six arrangements). Three heads/one tail and three tails/one head are also fairly likely (four arrangements each), while all heads or all tails has a much lower statistical entropy, since there is only one possible arrangement.

Entropy can also be a measure of the extent to which a collection of objects (atoms in a gas or liquid, stars in a galaxy, files on a hard drive, etc.) is disordered. This is statistical entropy, where the likelihood of a specific state occurring by chance is considered—if you flip 100 coins, you might wind up with 100 heads, but this is extremely unlikely as there is only one possible arrangement (low entropy). You are far more likely to get 99 heads and one tail (100 possible arrangements), and the most likely (highest entropy) result is 50 of each (large numbers of possible arrangements: 10^{29} of them, in fact).

light

ᴥ R.G.B. (red, green, blue)

the human eye contains three different types of cone-shaped light receptors, which detect three different ranges of wavelengths. Other animals see other ranges of wavelength, and can have different numbers of "cones." Different colors trigger different combinations of cones (e.g., red triggers only the yellow-sensitive cones, yellow triggers a mixture of yellow- and green-sensitive cones), allowing a full spectrum of colors to be detected. A similar system of mixing three colors of light is used for computer and T.V. screens. An intensity is assigned to each of red, green and blue, which combine to give the desired color (black is no illumination at all, white is full brightness from all three). Most current computer graphics systems use 8 bits for the intensity of each of the three colors, allowing 16.7 million colors (though specialist hardware and software often use more bits per color). Both systems are additive mixing—you start with darkness, and add color and light.

Humans can detect the range of light spectrum from about 400nm to about 700nm. We see this as a smoothly varying rainbow of colors, from violet (higher frequency) to red (lower frequency).

C.M.Y.K. (cyan, magenta, yellow, key (black))

this is the color model used by printers and is similar in principle to the red, blue and yellow paint mixing taught to children, though the different primary colors used in printing give better clarity of color. The system is subtractive: the more of each color added, the less light is reflected. Small dots of ink are printed in each color (the size of the dot varies to give different intensity of that color), and from a distance the dots blur to give the appearance of a single color. Black is added as a separate color because, while in theory combining all the primaries at maximum intensity should give black, the practical result tends to be murky brown with some of the paper showing through.

ᴥ H.S.V. (hue, saturation, value)

a system used mainly by artists to mix opaque paints, though it can be used to describe colors in other media, including some computer graphics programs.

hue

the element of a color that most people would think of as "color." It is usually described as a continuous wheel, with the spectrum along the rim, and red shading into purple.

chroma (or saturation)

the vibrancy, or intensity, of a color. A low-intensity color will be a shade of gray with a hint of the **hue** in it. Pastels are very

A standard rrepresentation of the H.S.V. color space. Hue = red, green or blue; Saturation = from achromatic white to pure hue; Value = brightness, from black to pure hue.

Physical Sciences Light

high-**value**. low-intensity colors. Measured as a percentage. 100 percent is a fully saturated color. and 0 percent is a pure gray. (Differs from **luminous intensity**.)

value (V)

a measure of how light or dark a color is (also sometimes called its brightness). Like **chroma**. measured as a percentage: pure black is V0 percent. regardless of chroma: pure white is V100 percent. regardless of **hue**. Some hues are perceived as intrinsically brighter than others and do reflect more light even at maximum saturation and value—yellow is the brightest color. and blue tones tend to appear the darkest.

luminous intensity

a measure of the energy emitted as light in a particular direction. A related but different measure is the radiant intensity. which is the amount of energy emitted as electromagnetic radiation (including light) in a particular direction. Because the eye responds differently to different colors of light. the same radiant intensity produces different luminous intensities for different wavelengths of visible light.

candle

an old name for the **candela**. To confuse matters. it is also an older unit of **luminous intensity**. which was replaced by the unit that was then renamed to become the candela. There were numerous local definitions of a standard candle, varying from around 0.9 cd to 1.1 cd. but a standard international candle was agreed in 1909. with a value about 2 percent greater than the modern candela.

candlepower

the output of a light source in units of **luminous intensity**. Formerly measured in standard international **candles**. and now defined using the **candela**—the two terms are essentially interchangeable but a "million-candlepower lamp" is more immediately understandable than "a lamp with an output of one million candela."

candela (cd)

the SI unit of luminous intensity. It is defined as the luminous intensity (in a given direction) of an object emitting (in the direction specified) $1/683$ watts per steradian of monochromatic electromagnetic radiation. with a frequency of 540 terahertz. The frequency is that at which the human eye is most sensitive (a slightly yellowish green). and the slightly random-seeming $1/683$ is to maintain the value produced by a previous definition involving a "black body" held at the melting point of platinum (around 1.770°C).

lumen (lm)

the SI unit of luminous flux. The total amount of light emitted into a solid angle of 1 **steradian** from a light source with a uniform (that is, equally in every direction) intensity is 1 **candela**. This is not the same as the energy flux (or power, in watts), which will vary for constant luminous flux depending on the wavelengths of light used. A lumen is defined by angle, not by area, and is still one lumen whether it is spread over one square meter or one square light-year.

lambert

a unit of luminance, the **luminous intensity** of a surface, i.e., the amount of light that the surface emits or reflects per unit area. A surface with luminance 1 lambert emits (or reflects) 1 lumen per square centimeter, which is calculated as equal to an illumination of over 3,000 **candela** per square meter. There is no standard abbreviation for the lambert, nor an equivalent SI unit, though the **lux** has the same basic dimensions.

1 meter 4,000 lux   2 meters 1,000 lux   3 meters 444 lux

As you move farther from a light source, the intensity of the light you receive from it will decrease.

lux (lx)

the SI unit of illumination of a surface, i.e., the light falling onto the surface. It is defined as 1 lumen per square meter. The illumination provided by most artificial light sources is extremely low, relative to natural sunlight—an office brightly lit with strip lamps will have an illumination of between 300 and 500 lux, while daylight ranges from around 30,000 lux (on a cloudy day) to over 100,000 lux.

foot-candle (fc)

a now little-used unit of illumination, defined as the illumination cast by a light source of brightness one international **candle** onto a surface 1 foot away. Later redefined as 1 lumen per square foot, making it equal to about 10.77 lux.

diopter

a measure of the focusing power of a lens, equal to one divided by the focal length of the lens. For a converging lens (which includes most everyday uses), this is a positive number; for a diverging one, the number is negative.

polarization

the angle at which a **photon**'s electrical component is oriented. While it cannot be measured for an individual photon, the polarization of a light source can be established by measuring the angle at which a polarizing filter allows the maximum amount of light through. Most normal light sources are not polarized (their individual photons have random polarizations), but light can become polarized by passing through a filter or reflecting from a surface. Polaroid (and similar) sunglasses work on this principle, by absorbing light polarized horizontally, to reduce glare.

reflectivity

the ratio of energy reflected to energy passed through for light hitting a surface. Reflectivity is extremely important for fiber-optic communications, since the laser signals used to transmit information will be reflected from the side of the fiber many thousands of times over even a short distance. If the proportion of the energy lost through the surface of the fiber (rather than reflected) is not extremely low, the signal rapidly becomes difficult to detect.

The refractive index of a substance controls how widely separated are the red and violet fringes produced by a prism of that material. (The same effect produces rainbows in the sky.)

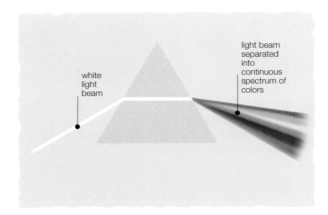

white light beam

light beam separated into continuous spectrum of colors

≳ refractive index (n)

a measure of the amount by which electromagnetic radiation (including light) is slowed down when traveling through a material relative to traveling in a vacuum. Calculated by dividing c (the speed of light in a vacuum) by the speed of light in the material in question, usually giving a number between one and three, though values over two are rare. The refractive index of a material also controls the amount by which light passing into it at an angle will bend; this is because the frequency of the light remains the same, but its wavelength changes (explaining why water always looks shallower than it actually is).

diffraction

the ability of a wave, including light, to bend as it passes around the edge of a barrier. The extent to which a given wave will diffract depends on the relationship between the width of the gap through which the wave passed and its wavelength, peaking when the two are approximately equal. Careful calculation of likely behavior in particular circumstances allows control of unwanted diffraction, whether of laser light in experimental apparatus or water waves around the entrance to a harbor.

optical density

a measure of an object's ability to absorb light. Each unit of optical density represents an order of magnitude decrease in the light passing through. That is, an object with an optical density of one allows 10 percent of light entering to pass through (absorbing the other 90 percent), while an optical density of two allows only 1 percent of light entering to pass through (absorbing 99 percent), etc. This is a cumulative, rather than per unit, measure, therefore if three objects each with optical density of one are placed in line, the total optical density is three and only 0.1 percent of the light passes through.

ꙮ laser

an acronym for Light Amplification by Stimulated Emission of Radiation. A device designed to produce highly focused, usually monochromatic, light, often in very short bursts. Since the beam of a laser is highly collimated (the light is all traveling in almost exactly the same direction), it cannot normally be seen unless an object is placed in the path of the beam to reflect it, which makes high-power laboratory lasers, able to cut through sheet metal (or a human being), extremely dangerous. The power of lasers in everyday use (primarily pointers and C.D. players) are typically measured in milliwatts. However, even these will cause permanent damage if pointed into an eye, and industrial lasers can have a power output up to a few tens of kilowatts. Lasers are very useful for making measurements related to a variety of optical properties, because the light produced can be very tightly controlled.

Lasers are extremely useful to scientists but invisible to anyone not directly in their beam of light.

wavelength

the distance between two adjacent peaks of a wave (including light, sound and water waves), equal to the speed at which it travels divided by its frequency. For light, this is typically in the range of 400 nanometers (violet) to 740 nanometers (red). A monochromatic light source is one that produces light only on a single wavelength. It is not possible to produce perfectly monochromatic light, but most lasers (and some other devices) come close enough for practical purposes.

*NASA has described the
Danjon scale in the precise
following terms:*
*0 very dark eclipse; moon
almost invisible, especially
at mid-totality*
*1 dark eclipse, gray or
brownish in coloration;
details distinguishable only
with difficulty*
*2 deep red or rust-colored
eclipse; very dark central
shadow, while outer edge of
umbra is relatively bright*
*3 brick-red eclipse; umbral
shadow usually has a bright
or yellow rim*
*4 very bright copper-red or
orange eclipse; umbral
shadow has a bluish, very
bright rim*

wavenumber

a measure of the properties of a wave, including light waves. The wavenumber (for a specified unit distance) is equal to one divided by the wavelength (measured in that unit). Wavenumbers for many types of useful radiation are often given in waves per centimeter. Visible light, for example, has between 14,000 and 25,000 waves per centimeter.

photon

the quantum (smallest possible quantity) of light or any other electromagnetic radiation. The energy of a photon in **joules** is equal to its frequency (in **hertz**) multiplied by the Planck constant (6.6×10^{-34}), and the laws of physics dictate that it remains in constant motion at the speed of light until its energy is absorbed and converted to another form—at which point the photon, which has no rest mass, ceases to exist.

rayleigh (R)

a very small unit of light intensity, primarily used for astronomy. Equal to one million photons per square centimeter per second.

❧ Danjon scale

a measure of lunar eclipse brightness, defined by the French astronomer André-Louis Danjon (1890–1967) on a scale from 0 to 4 (*see* caption).

finsen unit (FU)

a measure of intensity for ultraviolet light, originally intended for medical use, but also relevant to suntanning, now essentially replaced by the **standard erythemal dose** (though the two measure slightly different things). One FU is the intensity equal to 10 microwatts per square meter for ultraviolet light of wavelength 296.7 nm (frequency 1.01 petahertz).

standard erythemal dose (S.E.D.)

a measure of cumulative absorbed energy from ultraviolet light, defined as 100 joules per square meter (normally of skin surface). The related measure of intensity is expressed in S.E.D.s per hour, equal to 27.8 milliwatts of skin-affecting ultraviolet light per square meter.

mathematics

gradient

a measurement of rate of change. Colloquially, the gradient is a measure of the steepness of a hill, and is defined in terms of a ratio. A 1:2 gradient increases or decreases one unit in height for every two units in length. In mathematical terms, a gradient is a **vector** acting on a scalar function at a given point in a scalar field. It may therefore exist in more than two dimensions.

apex

the highest point in a two-dimensional plane or three-dimensional solid figure, relative to a base line or lowest point. Most commonly, the word apex is used to describe the tip of a cone, or more specifically the end of the axis opposite the base. Colloquially, the apex of an object is simply its highest point, such as the apex of a roof or the apex of a church spire.

vertex

a mathematical term having a number of similar meanings. Geometrically, a vertex is any of the angular points of a polygon or polyhedron. In this sense it sometimes used interchangeably with **apex**, but strictly speaking the apex is only one of the vertexes. A vertex is also the point at which an **axis** meets a curve. Colloquially, the vertex of an object is also used interchangeably with apex to mean the highest point.

axis

an imaginary straight line. In geometry, an axis is the imaginary straight line around which a plane is rotated to create a symmetrical solid. In a wider mathematical sense, an axis is a fixed line against which coordinates are measured, commonly named an x-axis, y-axis or z-axis. An axis of symmetry is the line in respect of which a two- or three-dimensional shape may be symmetrical.

❧ point of inflexion (inflection)

the point at which a curve changes from being concave to convex. More specifically, a point of inflexion is the point at which the sign (positive or negative) of the rate of change of the gradient itself changes.

asymptote

a straight line that becomes continually closer to a particular curve, without ever meeting that curve. Asymptotes are

The point of inflexion for this curve is that point where the gradient of the curve changes from being positive to negative.

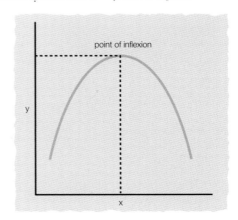

graphical examples of Zeno's dichotomy paradox. If a man is standing a certain distance from a wall, and he continually divides the distance between him and the wall by half, he will never reach the wall—there will always be a certain distance between him and the wall even if it is imperceptibly small. In the same way, a curve will never touch its asymptote, even though the distance between them becomes infinitesimally minute.

❧ arc degree (°)

a unit of angular measure, also simply known as a degree. There are 360° in a full circle, 180° in a semicircle, and 90° in a right angle. Arc degrees are ways of measuring the size of angles, but can also be used to measure lengths in relation to distance. For example, your thumb, held at arm's length from you, has a length of about 2°. Arc degrees are subdivided into arc minutes and arc seconds.

arc minute (′)

a subdivision of the arc degree. There are 60 arc minutes (60′) in an arc degree (1°). Arc minutes are subdivided into arc seconds.

arc second (″)

a subdivision of the arc minute. There are 60 seconds (60″) in an arc minute (1′), and 3,600 arc seconds in an arc degree (1°).

radian (rad)

the SI approved unit of angular measure, and an alternative to the **arc degree**. One radian is the angle subtended at the center of a circle by an arc of the same length as the radius of that circle. As such, there are 2π radians in a circle, and 1 radian is equivalent to approximately 57.3°. Radians are used in calculus to make results as natural as possible.

❧ steradian (sr)

a unit of three-dimensional angular measure. The steradian is the SI unit of solid angle, and the three-dimensional equivalent of the **radian**—for this reason it is sometimes also called a square radian. One steradian is the angle subtended at the center of a sphere by a part of the sphere whose surface area is equivalent to the square of the radius of the sphere.

segment

a line segment is that part of a line included between two points. When both points lie on a circle, the line

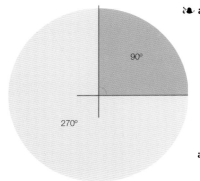

A circle is divided into 360 arc degrees, 32,400 arc minutes and 1,944,000 arc seconds.

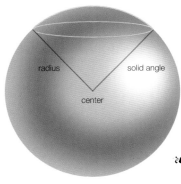

Steradians, being measures of three-dimensional angles, are always conical in shape. There are 4π steradians in a sphere.

segment is a chord. A circle segment is that part of a circle enclosed by a chord and an arc. The segment is the smaller of the two portions thus created, or the portion that does not contain the circle's center point. A sphere segment is that portion of sphere cut off by the intersection of any plane that does not pass through its center.

sector

the portion of a circle whose boundaries are two radii of the circle or ellipse and the arc between them. A sector is also a mathematical implement. Consisting of two hinged arms, it is marked with **sines** and **tangents** for the purpose of creating diagrams.

sine

for one of the non-right-angle corners of a right-angle triangle, the sine is the length of the side opposite the angle as a fraction of the hypotenuse. The equivalent for the side adjacent to the angle is the cosine.

✺ tangent (tan)

a mathematical term which has two distinct meanings. In geometry, a tangent is a straight line which touches a curve at a point where the curve and the straight line have the same gradient. The geometrical tangent of a curve is defined using calculus. In trigonometry, the tangent is the ratio of the sides other than the hypotenuse opposite and adjacent to an angle in a right-angled triangle. The tangent is defined sin x/cos x.

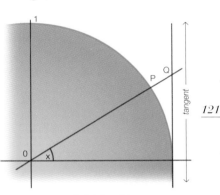

Definition of the tangent function. The unit circle is the circle with its center at the origin and a radius of 1. Angle x is formed by rotating OA about the origin to OP. Point Q is the intersection of line OP and x = 1. Then the y-coordinate of point Q is tan x.

integer

a whole number. The set of integers is that set of numbers which includes all the positive natural numbers (1, 2, 3, 4 ...), all the negative natural numbers (−1,−2, −3, −4 ...) and 0. Any number containing a fraction or a decimal is not an integer. The set of integers is indicated by mathematicians by the letter **Z**, which stands for *Zahlen*, the German for numbers. Number theory is the branch of mathematics that includes the study of integers.

real

a word used to describe any number that contains no **imaginary** part; indeed the term real was coined by mathematicians in response to the concept of imaginary numbers. In practical terms, a real number is any number that corresponds to a point on the infinite number line. Real numbers may be positive, negative, rational, irrational, algebraic or zero.

imaginary

the word used for the non-real part of a complex number that can be written in the form a + ib, where a and b are real numbers and **i** is the square root of −1. A complex number is one that has both real and imaginary parts—"ib" is the imaginary number. The concept of the imaginary number was coined by Frenchman René Descartes in his 1637 work *La Géometrie*. His assumption was that such numbers did not exist—hence imaginary. In fact, imaginary numbers are central to many sciences and branches of advanced math.

prime

any natural number whose only positive divisors are itself and 1. In number theory, prime numbers are the basic building blocks of all natural numbers, which is to say that any integer can be expressed as the product of prime numbers. There are an infinite number of prime numbers, but the largest known prime is $2^{24,036,583} - 1$, which has 7,235,733 digits. The opposite of a prime number is a composite number.

rational

a word used to describe any number that can be expressed in terms of the ratio of two integers. In other words, a rational number is any number that can be expressed in the form a/b, where a and b are integers and b does not equal 0. Hence, 2 is a rational number because it can be expressed in the form $8/4$, or $16/8$ or any other number of such ratios. Similarly $1/2$ is a rational number because it can be expressed in the form $2/4$, or $4/8$, and so on. The opposite of a rational number is an **irrational** number.

irrational

a word used to describe any real number that cannot be expressed in terms of the ratio of two integers. In other words, an irrational number cannot be expressed in the form a/b, where a and b are integers and b does not equal 0. Examples of irrational numbers are the square root of 2, π and **e**. The opposite of an irrational number is a **rational** number.

infinity (∞)

a number greater than any assignable value. The concept of infinity has evolved over history, and has philosophical and cosmological as well as mathematical connotations.

i

an imaginary number equal to the square root of −1. The mathematical symbol i stands for imaginary unit, and it is properly defined as the solution to the equation $x^2 = -1$.

e

a mathematical constant equal to approximately 2.71828. The mathematical symbol e stands for Euler's number, and is named after the Swiss Leonhard Euler (1707–83). It is a transcendental number, and the base of the natural **logarithm** function.

❧ pi (π)

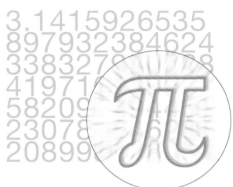

a mathematical constant approximately equal to 3.1415927. Pi is defined as the value of the ratio of the circumference of a circle to its diameter. It is irrational, transcendental and non-constructible. Pi is one of the most important mathematical constants, essential to describing circles: the area of a circle is πr^2, where r is the radius; and the circumference of a circle is $2\pi r$. The Greek letter π used to represent this constant is the initial letter of the Greek word *periphereia*, meaning circumference. As pi is an irrational number, throughout history much effort has been applied to calculating it to the highest number of decimal places. The current record is 1,240,000,000,000 decimal places.

Pi is perhaps the most important constant in mathematics. It is an irrational number, which means it has an infinite number of decimal places.

countable

a word used to describe the contents of a set when the number of elements can be counted. A set with more elements than is countable is uncountable, but there are uncountable sets of different sizes. A countable set may be infinite, e.g., the set of integers.

random

a statistical term used to describe events that happen in such a way that cannot be accurately predicted. This is not to say, however, that random events are unpredictable when taken en masse. For example, snowflakes may fall randomly, but it is possible to predict the general area in which they will fall, and the cumulative nature of their falling. In mathematical terms, randomness gives rise to statistical theories of **probability**, a means of describing the probable outcomes of a large number of random events.

arithmetic sequence

a finite or infinite sequence of numbers in which each number is equal to the previous number plus a constant, also known as the "common difference." So, an arithmetic sequence starting with 3 and with the common difference 4 would begin 3, 7, 11, 15, 19 … Arithmetic sequences are notated algebraically when the first element is a 0 and the common difference is d: $a_n = dn + a_0$.

geometric sequence

a finite or infinite sequence of numbers in which each number is equal to the previous number multiplied by a constant, also known as the "common ratio." So, a geometric sequence starting with 3 and with the common ratio 4 would begin 3, 12, 48, 192, 768 … Geometric sequences are notated algebraically when the first term is a_0 and the common ratio is r: $an = a_0 r^n$.

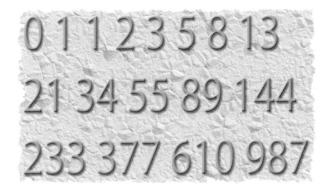

exponential sequence

a finite or infinite sequence of numbers in which each number is equal to the previous number raised to the power of a constant exponent. So, an exponential sequence starting with 3 and with the exponent 4 would begin 3, 81, 43046721 ... Exponential sequences are notated algebraically when the first term is a_0 and the exponent is x: $a_n = a_0^{\left(x^n\right)}$.

ఎ Fibonacci sequence

an infinite sequence of numbers in which each number is equal to the sum of the previous two numbers. The simplest Fibonacci sequence is that which starts with the numbers 0 and 1, and begins 0, 1, 1, 2, 3, 5, 8, 13 ...

average—mean

an average figure derived from taking the sum of several quantities and dividing it by the number of quantities. Thus, the mean average of 1, 3, 5 and 7 is $(1+3+5+7)/4 = 4$. The arithmetic mean is often used, but it can be very misleading, as it can be skewed by unrepresentatively large or small figures at either end of the population. So, if five people earn $10,000, $10,000, $13,000, $15,000 and $150,000, the mean average of their incomes is $39,600, which says nothing meaningful about the average income.

average—mode

an average figure equal to the most common quantity in a number of quantities. For example, the modal average of 1, 2, 2, 3, 3, 3, 4, 5 is 3, as this is the figure that appears the most often. There need not always be a modal average in a quantity of numbers, if two or more numbers appear equally frequently.

average—median

an average figure below which 50 percent of the quantities in a population fall. If there is an odd number of quantities in the population, the median will be a number in the population. If there is an even number of quantities in the population, the median will be the mean average of the two central quantities in the population. So, the median average of 1, 2, 2, 3, 5, 6, 8, 8, 9, 10 is 5.5, the figure below and above which 50 percent of the figures fall.

average—midrange

the mean average of the highest and lowest quantities in a number of quantities. For example, the midrange average of 1, 3, 5, 6, 6, 6, 7, 11 is (11+1)/2 = 6.

☙ probability

a means of estimating the likelihood of the occurrence of an event. In mathematics, probability is generally represented in terms of a real number between 0 and 1. An impossible event has probability 0, and a certain event has probability 1. Thus a rare event has a probability close to 0, whereas a common event has a probability close to 1. More colloquially, probability is represented in terms of fractions: the probability of a dice roll being 3, 5 or 6 is $^1/_2$ (0.5). Gamblers represent probability in terms of odds, or ratios: a 2:1 chance of an event occurring is the equivalent of a $^2/_3$ probability.

permutation

the number of different ways in which a combination of numbers may be ordered. So, the combination of numbers {1, 2, 3} has six permutations: {1, 2, 3}, {1, 3, 2}, {2, 1, 3}, {2, 3, 1}, {3, 1, 2}, {3, 2, 1}. The number of permutations of a **combination** is equal to the **factorial** of the number of elements in that combination.

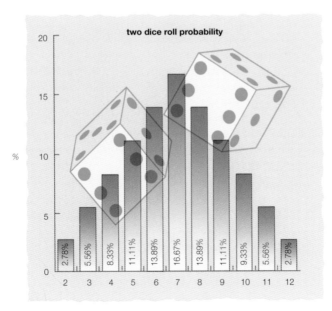

The probability of rolling a three is 1/6, but that does not mean that if you roll a dice six times you will definitely roll a 3. The probability of a given event is not affected by previous events. This graph illustrates the probability of getting a particular number when you roll two dice.

combination

a selection of a number of elements from a larger number of elements, without regard to the way in which those are ordered. For example, three combinations of three numbers may be derived from the sequence {1, 2, 3, 4}, namely {1, 2, 3}, {1, 2, 4} and {2, 3, 4}. However, there are six **permutations** of each of these combinations.

googol

a number equivalent to 10^{100}, or 1 followed by 100 zeros. The term was "invented" by a nine-year-old in 1938, Milton Sirotta, the nephew of American mathematician Edward Kasner (1878–1955). The googol does not have any particular mathematical use, though it is interesting to note that a regular polygon with a googol number of sides would appear as a circle even at the scale of a **Planck length**. It is believed that the number of particles in the known universe is fewer than a googol. The name of the Internet search engine "Google" is a pun on the word googol.

googolplex

a number equivalent to 1 followed by a **googol** of zeros, or ten raised to the power of a googol. If the digits of a googolplex were written in 1-**point** type, its length would be 4.7×10^{69} times the diameter of the known universe.

standard deviation (σ; s)

a measure of statistical dispersion. The standard deviation of a set of numbers is the mean difference between those numbers and their mean. The mean of the numbers 1, 3, 4, 6, 7 is 4.2, and the standard deviation is approximately 2.387, indicating that on average the difference between each number and 4.2 is 2.387. A large standard deviation tells us that on average the elements of a population are a long way from the mean. For example, the populations 29, 30, 31 and 20, 30, 40 both have a mean of 30, but the standard deviation of the latter is much larger than that of the former.

normalize

the process of multiplying a series, function or number by a factor such that some associated quantity—often, but not necessarily, the norm—is equal to a desired value, usually 1.

quantiles

equal divisions of a frequency distribution of a random variable, such as **quartiles** or other **percentiles**. Quantiles are represented graphically by vertical lines dividing the distribution graph of the variable.

quartiles

four equal groups into which a population of values of a particular variable can be divided. Each quartile represents a quarter of the population. The first quartile cuts off the lowest quarter of the data, and is also known as the 25th percentile. The second quartile cuts the set of data in half, and is also known as the median or the 50th percentile. The third quartile cuts off the highest quarter of the data, and is also known as the 75th percentile. The difference between the first and third quartile is known as the interquartile range.

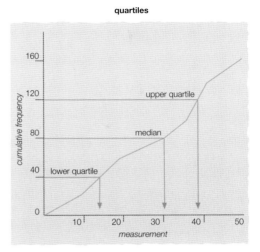

percentiles

a hundred equal groups into which a population of values of a particular variable can be divided. The 25th percentile, the 50th percentile and the 75th percentile are known as **quartiles**. If a figure appears in the 95th percentile, then it is in the top 5 percent of that population.

percent (%)

a means of representing a decimal, fraction or proportion as a whole number. Percent derives from the pseudo-Latin phrase *per centum*, meaning out of a hundred, hence 25 percent represents the fraction $25/100$. Percentages can be greater than 100 percent—150 percent, for example, represents an increase of 50 percent. Put another way, 150 percent = $150/100$ or 1.5.

Quartiles are a type of percentile: the 25th or lower percentile, the 50th (median) percentile and the 75th or upper percentile.

numerator

the number above the line in a vulgar fraction. The numerator indicates how many of the parts indicated by the **denominator** the value of the fraction equals. So, for example, in the fraction $2/3$, the numerator is 2, and it indicates that the fraction is equal to 2 one-third parts.

denominator

the number below the line in a vulgar fraction. The denominator indicates the number of parts into which the population the fraction describes is divided. So, in the fraction $3/4$, the denominator is 4, and it tells us that the population is divided into four equal parts, or quarters. In $3/4$ the **numerator** 3 indicates that the fraction is equal to 3 of the 4 quarters. Denominators are never equal to 0.

The numerator is the number on the top of a vulgar fraction, in this case 17; the denominator is the number on the bottom of a vulgar fraction, in this case 45.

factors

an integer by which another number is exactly divisible. For example, 1, 2, 3 and 6 are all factors of 6, because 6 can be divided precisely by any of them. Factors can be expressed algebraically. For example the factors of 6x are 1, 2, 3, 6, x, 1x, 2x, 3x and 6x.

factor

a number by which another number is increased or multiplied. So, 10 increased by a factor of 4 is 40.

factorial (!)

the product of a positive integer and all the integers below it except 0. So, 6! is equal to $6 \times 5 \times 4 \times 3 \times 2 \times 1 = 720$. 0! is taken to be equal to 1. Factorials are used in working out **combinations** and **permutations**. There are n! permutations of n objects. The number of combinations of choosing k objects from n objects is equal to n!/k!(n − k)!. Factorials are also used widely in calculus and probability theory.

bases

a number used as the basis of a scale of counting (e.g., **binary** (base 2), **decimal** (base 10), etc.). A particular base n uses the digits between 0 and n − 1: so base five uses the digits 0, 1, 2, 3 and 4. The base of a **logarithm** system is a number to which all numbers in that system are referred.

binary

base 2. The binary system of numeration uses only the digits 0 and 1. The decimal number 2 is rendered as 10 in binary; the decimal number 5 is rendered 101. The binary system is fundamental to Boolean logic and hence to the operation of modern electrical circuits and computers, because the two digits can represent two different voltages.

decimal (denary)

base 10. Decimal is the most commonly used system of numeration and uses the 10 digits 0, 1, 2, 3, 4, 5, 6, 7, 8 and 9. It is widely assumed that humans have mostly adopted the decimal system of counting because they have 10 fingers and 10 toes. Not all human civilizations have used decimal, however. Base 8, or octal, has been used by the Maya, the Babylonians and the Yuki Indians.

hexadecimal

base 16. Hexadecimal uses the following 16 digits: 0, 1, 2, 3, 4, 5, 6, 7, 8, 9, A, B, C, D, E, F. In hexadecimal, the decimal number 10 is written A; the decimal number 16 is written 10; the decimal number 100 is written 64; and the decimal number 1,000 is written 3E8. Hexadecimal is used by computers. because four **binary** numbers. or bits. can be represented easily as a single hexadecimal digit. For example. 1011 in binary is equal to B in hexadecimal.

power

the result of a number being multiplied by itself a certain number of times. Hence. 4 to the power of 6. or 4^6. is equal to $4 \times 4 \times 4 \times 4 \times 4 \times 4 = 4.096$. In this example. 4 is known as the **base**. and 6 is known as the exponent. The exponents 2 and 3 are used so often that they have more common names: the square and the cube respectively. Negative exponents have the effect of reciprocating the power. so 4^{-6} is equal to $1/4^6 = 1/4096$. The exponent 1 is seldom written. as $a^1 = a$. Any number raised to the power of 0 is defined as being equal to 1.

exponential

a function in which a base number is raised by a certain **power**. or exponent. to produce a given result. The inverse of an exponential function is a logarithmic function.

❧ logarithm (log)

the power to which a base number must be raised to produce a given result. The logarithmic function is therefore the inverse of the exponential function. Logarithms are represented thus: $\log x^a = b$ where $x^b = a$. So. $\log_8 64 = 2$. because $8^2 = 64$. However. $\log_4 64 = 3$ because $4^3 = 64$. Logarithms are often used to solve equations where the exponent is unknown. and they often appear in calculus. especially as the solutions to differential equations. Natural logarithms are logarithms whose base is **e**.

Slide rules were traditionally used to calculate logarithmic functions. They were a mechanized form of logarithmic tables that did not require the user to know the logs. They have been all but superseded by calculators and computers.

order of magnitude

a scale of size. typically determined by powers of 10. So. if a number b is 1 order of magnitude bigger than a. it is 10 times bigger. If it is 2 orders of magnitude bigger than a. it is 100 times bigger. and so on. If the numbers a and b are of the same order of magnitude. it does not mean that they are the same. rather that the difference between them is less than 10 times. An order of magnitude. therefore. is an approximate position on a logarithmic scale of **base** 10.

❧ golden ratio (φ)

a ratio, traditionally thought to be particularly pleasing and which occurs in mathematics, geometry and nature with some regularity. Written as a number, the golden ratio is approximately equal to 1.618033, and is often represented by the Greek letter phi (φ). The ratio of a regular pentagon's sides to its diameter is equal to the golden ratio, and it is used to describe the **Fibonacci sequence**. Paper sizes used to be based on the golden ratio, although the current standard sizes are based on √2.

The diameter of a pentagon (the distance from one vertex to the center of the side opposite) is equal to the length of one side multiplied by the golden ratio.

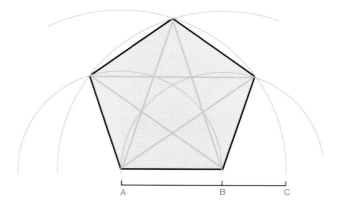

scalar

a quantity that is described by magnitude but not direction. Scalars, therefore, are distinct from **vectors**. Distance is an example of a scalar measurement; if we say that an object is 100 meters away, we know nothing about the direction we need to travel in order to reach that object.

vector

a quantity that is described by both magnitude and direction. Vectors, therefore, are distinct from **scalars**. Velocity is an example of a vector measurement as it measures the rate and direction of motion. In advanced mathematics and physics, vectors can be more than two-dimensional. Spacetime, for example, is described in terms of four-dimensional vectors, or four-vectors.

orthogonal

at, or pertaining to, right angles. A line that is orthogonal to another is at right angles, or perpendicular, to it.

nuclear and atomic physics

quark flavors

a property by which a quark is defined. There are six flavors of quark: up (u), down (d), charm (c), strange (s), top (t) and bottom (b). Quarks are subatomic particles whose existence has been suggested experimentally, but which have not actually been observed. As such, traditional means of measuring and describing them are not useful. For this reason, they are defined by their flavor and their color.

atomic number (Z)

the number of protons found in the nucleus of an atom. All elements possess different atomic numbers, and the atomic number of an element determines its place on the **periodic table** of elements.

elementary charge (e; q)

a unit of electric charge in the system of atomic units. It is equivalent to the negative electrical charge carried by a single electron, or the positive electrical charge carried by a single proton. It is about $1.6021764\underline{6} \times 10^{-19}$ coulombs. Elementary charge was first measured by the American physicist Robert Millikan (1868–1953) in 1909, and is believed to be the smallest unit of electric charge.

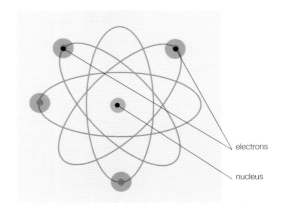

For this atom—which has four electrons—to be a "neutral" atom (with no overall electrical charge), the nucleus would need to contain four protons. If the nucleus contained six protons, it would be a positive ion, with a charge of +2e.

electrons

nucleus

electron mass

the mass of an electron. It is equivalent to approximately 9.1094×10^{-28} grams.

rest mass

in general usage, the mass of an object. The rest mass, or invariant mass, is distinct from the relativistic mass, a term used in special relativity to describe mass that changes according to one's frame of reference—for example, the relativistic mass of an object increases with its velocity. Rest mass, by contrast, does not change, and all observers of an object can agree on what the rest mass is. Relativistic mass became used less and less with the introduction of general relativity and, latterly, quantum mechanics, to the extent that physicists almost always mean "rest mass" when they use the word "mass."

੩ de Broglie wavelength (λ)

Every object has an associated wavelength; the smaller the object, the longer its de Broglie wavelength—but even subatomic particles have very short wavelengths.

the wavelength of a particle. A wavelength is the distance between the repeated crests of a wave pattern. It is named after the French physicist Louis de Broglie (1892–1987) who discovered that any particle that has momentum also has a wavelength—this discovery earned him the Nobel Prize for Physics in 1929. The de Broglie (it is pronounced "de Broy") wavelength of a relativistic particle can be calculated as being equal to h/p. where h is the Planck constant and p is the particle's momentum.

atomic mass unit (amu)

a unit of mass used for expressing atomic and molecular masses. It is also called the unified atomic mass unit (u) or the Dalton (Da). One atomic mass unit is equal to one-twelfth of the mass of a carbon-12 atom. or 1.66×10^{-30} grams. The atomic mass unit is not an SI unit, though it is generally accepted under that system. One hydrogen atom has a mass of approximately one atomic mass unit.

magic number

one of seven atomic masses that occur frequently. The magic numbers are 2. 8. 20. 28. 50. 82 and 126. If an atom has a magic number of nucleons. it is more stable than an atom which has one more or one fewer nucleon.

relative atomic mass

also called atomic mass and atomic weight. the relative atomic mass of an atom is the number of nucleons (protons and neutrons combined) in that atom. The atoms of a particular element always have the same atomic number. but do not always have the same relative atomic mass.

packing fraction

a measure of the stability of an atom. The packing fraction is the difference between the relative atomic mass and the mass number of the atom. divided by the mass number. The packing fraction in another sense is the proportion of space within a crystal. or other solid. that is actually formed from the atoms of that crystal.

੩ Bohr radius

the radius of the lowest energy unit in the model of the hydrogen atom described by the Dane Niels Bohr (1885–1962). Bohr was a pioneer of quantum physics. and his theory of the structure of the atom was the first to use quantum theory. It led to his winning the Nobel Prize for Physics in 1922. The Bohr radius is used as a unit of length in atomic physics, and is equal to 5.292×10^{-11}, or about half an **angstrom**.

shed

a unit of area used in particle physics equal to 10^{-52} square meters. One shed is equal to 10^{-24} barns, and it used to describe the cross-sectional area of a particle from which other particles are scattered. The shed is not an SI unit, and such areas are more properly described in terms of square **femtometers** (fermi).

barn

a unit of area used in particle physics equal to 10^{-28} square meters. The word is thought to have derived from the phrase "as big as a barn door." The barn is not an SI unit, and such areas are more properly described in terms of square femtometers (**fermi**). One barn is equal to 100 square femtometers.

quantum

a unit of relative energy. The amount of energy in a quantum is proportional to the radiation it represents and is equal to the frequency multiplied by the Planck constant. An extended meaning of the word quantum is the smallest measurement into which certain other physical properties may be divided. **Elementary charge**, for example, is the quantum of electrical charge.

becquerel (Bq)

a unit of radioactivity. Named after French physicist Antoine-Henri Becquerel (1852–1908), who shared the 1903 Nobel Prize for Physics with his compatriots Marie and Pierre Curie for discovering natural radioactivity in uranium salts, the becquerel is the approved SI unit of radioactivity. Radioactivity is the disintegration of atoms; as they disintegrate they emit energy. One becquerel is equal to the radiation caused by the disintegration of 1 nucleus in 1 second. One becquerel is approximately equal to 27 picocuries (**curie**).

curie (Ci)

a unit of radioactivity. Originally defined as the radioactivity of 1 gram of radium-226, it was agreed in 1953 that 1 curie should be equal to 3.7×10^{10} atomic disintegrations per second, or 37 gigabecquerels. Named after Marie and Pierre Curie (1867–1934 and 1859–1906 respectively), the discoverers of radium, the curie is not an SI unit of measurement, having been superseded by the **becquerel**.

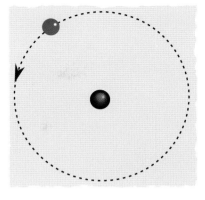

The classic representation of the hydrogen atom as shown by Niels Bohr.

sievert (Sv)

an SI unit measuring the effective dose of radiation received by a living organism. It is named after Swedish physicist Rolf Sievert (1896–1966). The sievert is equal to the actual dose of radiation multiplied by a factor which varies according to how dangerous the radiation is—the higher the factor, the more dangerous the radiation. The effective dose of natural background radiation is approximately 1.5 millisieverts a year. If the effective dose is raised between 2 and 5 sieverts it can cause hair loss, nausea and, in some cases, death. One sievert can be subdivided into 100 rem.

gray (Gy)

an SI unit of energy measuring the absorbed dose of radiation. It is named after English radiobiologist Louis Gray (1905–65). One gray is defined as being the dose of 1 joule of energy absorbed by 1 kg of matter. It is equal to 100 **rad**. The gray differs from the **sievert** in that it does not differentiate between less and more harmful types of radiation.

half-life

a unit of relative time denoting the time it takes for the radioactivity of a particular isotope to reduce by half. After a second half-life, the radioactivity of the isotope will again reduce by half, reducing it to a quarter of its original radioactivity. To reduce the radioactivity of a substance to 0.1 percent of its original level takes just under 10 half-lives. Half-lives vary wildly between elements. The half-life of uranium-238 is 4,500 million years; the half-life of radium-221, on the other hand, is 30 seconds.

half-value layer

the amount of material required to reduce the intensity of radiation by half. The half-value layer is a means of measuring the intensity of a radiation source. Different radiation sources have their half-value layers defined in terms of different materials. For example, the half-value layers of X-rays are generally described in terms of aluminum or copper thicknesses.

k factor

a measure of the strength of the gamma rays produced by a radioactive material. The k factor is measured in **roentgens** per hour at a distance of 1 cm from a source with a disintegration rate of 3.7×10^7 per second (or 1 millicurie).

kerma

the sum of the initial kinetic energies of all the charged particles liberated by uncharged ionizing radiation in a sample of matter, divided by the mass of that sample. Kerma is an acronym for kinetic energy released in materials. It is measured in **grays**.

roentgen (R)

a unit of ionizing radiation. It is named after German physicist Wilhelm Roentgen (1845–1923), who discovered the X-ray. As radiation hits an atom, it ionizes it by removing one or more electrons. This causes many of the biological effects of radiation, and the roentgen is a means of measuring these effects. One roentgen is the dose of radiation that liberates positive and negative charges of 2.58×10^{-4} coulombs of electric charge per kilogram of air.

pastille dose

a measure of radiation dosage. One pastille dose is the amount of radiation required to change the color of a pastille of barium platinocyanide from a particular green color to a particular red-brown color.

rad

a metric unit of radiation dosage. Rad is an acronym for radiation absorbed dose, and 1 rad is equal to 0.01 **gray**. In other words, 1 rad is equal to 0.01 joules of energy absorbed per kilogram of tissue.

rutherford (Rd)

a unit of radioactivity. One rutherford is equal to 1 mega-becquerel (**becquerel**), or one million radioactive disintegrations per second. The unit is named after New Zealand physicist Ernest Rutherford (1871–1937), who is commonly regarded as the founder of nuclear physics and the first person to suggest that the positive charge of an atom, and practically all its mass, are found in its nucleus.

ᔑ Hounsfield scale

a scale for measuring radiodensity, or how transparent a substance is to X-rays. A radiolucent substance is more transparent to

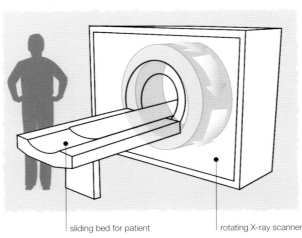

A CAT scanner uses the radiodensity scale devised by Godfrey Hounsfield to measure the internal anatomy of the body.

sliding bed for patient   rotating X-ray scanner

X-ray photons than a radiodense substance. and so has a lower score on the Hounsfield scale. The Hounsfield scale uses Hounsfield units (H.U.). Distilled water at standard temperature and pressure has 0 Hounsfield units: air has −1.000 Hounsfield units. The scale was established by British electrical engineer Sir Godfrey Hounsfield (1919–2004). one of the developers of computed axial tomography (commonly referred to as CAT scans). the principal use of which is imaging the internal anatomy of living organisms.

dollar ($)

a unit of reactivity of a nuclear reactor. One dollar is the level of re-activity in a nuclear reactor at which the chain reaction is just self-sustaining. It is not a fixed unit. however. varying as it does according to the size and build of the reactor.

inhour (ih or inhr)

a unit of reactivity of a nuclear reactor. The unit was first intro-duced in 1947 by Italian–American physicist Enrico Fermi (1901–1954). the creator of the first nuclear reactor. One inhour is equal to the amount of reactivity which will cause the number of neutrons in a reactor to increase by a factor of **e** in one hour. It is not a fixed unit. however. varying as it does according to the size and build of the reactor.

A dosimeter is usually shaped like a pen so that it can be easily clipped to the clothes. It can then measure the wearer's exposure to harmful environmental factors such as radiation or noise.

dosimeter

a measuring device used to measure exposure to a potentially harmful environment. A radiation dosimeter measures exposure to ionizing radia-tion. It resembles a pen. and is clipped onto the clothing where it measures how much radiation the individual is absorbing. Because radiation absorption is cumulative. the dosimeter should be worn every time the individual is in the presence of a radioactive source. and since it needs to be periodically recharged. the levels are generally logged so that users can keep track of their exposure levels. Dosimeters may also be used to measure exposure to other harmful environments besides radiation. A sound dosi-meter. for example. measures the user's exposure to harmful levels of sound.

Geiger counter

a device for measuring levels of radioactivity. Named after its creator. German physicist Hans Wilhelm Geiger (1882–1945). it detects and counts ionizing particles in the atmosphere. The Geiger counter can detect photons. alpha radiation. beta radiation and gamma radiation, but not protons. Some Geiger counters use a visual measuring device such as a needle: others produce an audible click. The Geiger counter has now been largely superseded by a new version of the device known as a halogen counter, which operates with a much lower voltage and has a longer life.

background radiation

radiation that occurs naturally in the Earth's atmosphere. This radiation is derived from a number of natural sources, some in the Earth itself, some in the atmosphere and some in space. That part of the background radiation which is left over from the Big Bang is called the cosmic microwave background and it is thought to exist throughout the universe. Background radiation varies, but it is generally in the range of 0.3–0.4 rem (see **sievert**)—a very low level of radiation which is believed not to be harmful.

critical mass

the minimum amount of fissionable material that will sustain a nuclear chain reaction. Various factors affect the critical mass of a material, including its nuclear properties, its shape, its purity and whether the material is surrounded by a neutron reflector. The shape that has the minimum critical mass is a sphere, meaning that less of a given material is needed to sustain a nuclear chain reaction if it is spherical. The critical mass for a sphere of uranium-235, for example, is 50 kg, compared to 15 kg for uranium-233. (These critical masses assume that the materials are not surrounded by a neutron reflector.)

❧ kiloton (kton)

a unit of energy equivalent to that which is released by the explosion of 1,000 tons of Trinitrotoluene (T.N.T.). This is roughly equal to 4.18×10^{12} joules, or 4 billion British thermal units. One thousand kilotons are a megaton. The kiloton is used to describe the destructive power of nuclear weapons—the bomb that was dropped on Hiroshima, for instance, produced about 15 kilotons of energy, while the largest nuclear weapon ever detonated produced 57 mega-tons of energy. If an asteroid measuring some $^{1}/_{2}$ mile (1 km) in dia-meter were to strike Earth, it is thought that it would produce more than 10 million megatons.

The nuclear bomb dropped on Hiroshima is thought to have had a destructive power of around 15 kilotons (more than 60 terajoules— enough energy to put five space shuttles into orbit).

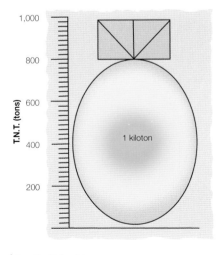

energy

the measure of the ability of an object (whether a photon of light, a tennis ball or an entire galaxy) to affect another object. It is often defined as the object's capacity for work, but this definition fails to take account of **entropy** (heat energy unavailable to perform work). Energy can take a wide variety of forms, though all can be classified as either kinetic (related to movement—this includes heat, which is the movement of atoms within a substance) or potential energy (stored energy—as, for example, in a chemical bond, coiled spring, or an object raised above the ground). Energy can be converted from one form to another, but can never be destroyed or created.

calorie (cal)

a unit of energy, defined as the amount of energy needed to raise the temperature of 1 gram of water by 1°C at a pressure of 1 atmosphere. Nutritional **Calories** (with a capital C) are 1,000 calories (with a small c), and are technically called kilogram-calories (although kilo calories is also acceptable) as they are the amount of energy needed to heat 1 kg of water by 1°C. As with the **Btu**, the temperature at which a calorie is calculated affects the precise energy value.

joule (J)

the SI unit of energy and work. It is the energy used to exert a force of 1 **newton** over a distance of 1 meter (and the work performed by doing so); the energy to move an electrical charge of 1 coulomb over a potential difference of 1 volt; or the energy to produce 1 watt of power for 1 second. One J is equal to 0.24 calories or just under $1/1000$ of a **Btu**.

kilowatt-hour (kWh)

the unit used by power companies to measure the amount of electricity used by a household. One kilowatt-hour is the amount of electricity that a 1 kW device uses in 1 hour (or a 2 kW device in half an hour), and is equal to exactly 3,600,000 J. Sometimes called a "unit" of electricity.

barrel of oil equivalent (boe)

a measure of the energy available from burning different types of fossil fuel, defined as 6.12 GJ, approximately the amount of energy released when one barrel of crude oil is burnt. This is around 6 million cubic feet (m.c.f.) of most types of natural gas, or 200 kg of coal; precise values vary depending on the nature and purity of the fuel being tested.

chemical energy

usually used simply to mean the **bond energy** of a chemical compound. It is occasionally used to describe the amount of energy released (or absorbed) during a chemical reaction, which is simply the total energy of all bonds created minus the total energy of all bonds destroyed, but this should more properly be termed the reaction energy.

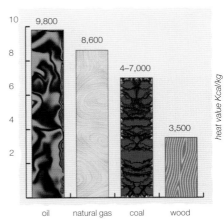

❧ heat of combustion

the energy released as heat when a material undergoes complete combustion—i.e., the substance being burnt reacts with as much oxygen as possible. For example, a fuel (such as methane or butane) consisting of a mixture of carbon and hydrogen has only undergone complete combustion when all the atoms in the fuel have reacted to form carbon dioxide and water. It can be measured per mole, per unit mass, or per unit volume of the material, and is usually used for comparing fuels (the higher the value, the better the fuel). It can also be used to help with fire safety equipment—the lower the heat of combustion, the less heat the substance will add to a fire. A substance with a negative heat of combustion actually absorbs heat as it burns.

Wood is a relatively poor fuel, in terms of energy released per kilogram, but has the great advantage that it is easy to grow more of it.

❧ calorimeter

a device used to measure the energy emitted (usually as heat) or absorbed by a chemical reaction or state-change (such as melting). The two most common kinds of calorimeter are the "bomb" calorimeter, used to measure energy changes in fast reactions, and the differential scanning calorimeter, which measures the change in energy over time, but many other specialist designs exist for various purposes.

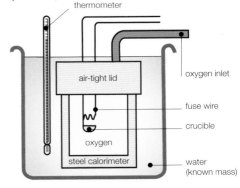

mass–energy conversion factor

according to Einstein's theory of special relativity, the mass of a body is a measure of the total energy contained within it (i.e., mass is simply another form of energy). This is the famous $E = mc^2$ where E is the energy equivalent to the rest mass m (the mass of the object when it is completely stationary). The energy released in nuclear reactions is due to this conversion: the products of fission or fusion have a slightly lower mass than the reactants.

A "bomb" calorimeter, for calculating the energy released by an explosive reaction or burning substance.

surface energy

the energy (usually measured in joules per square meter) required to break the chemical bonds within a substance and create a new surface. As there is an interaction between the atoms on each side of the surface. the actual value of the surface energy depends on both the substance and its surroundings—the surface energy between diamond and air is not the same as that between diamond and water. Generally. the greater the surface energy. the harder it is to break an object (though other factors may also come into play).

Energy can be stored in a variety of ways. While capacitors store electrical energy. batteries store chemical potential energy. which is only converted to electrical energy when they are connected to a circuit.

❧ potential energy (PE)

in effect. stored energy which an object can release to perform work (by moving to a state with lower energy). Types include gravitational PE (an object at a height can fall—PE = the mass of the object × the distance it can fall × the strength of the gravitational field). elastic PE (stretched or compressed materials can spring back). and electrical PE (which drives electrons round a circuit). Chemical **bond energy** is also a form of potential energy.

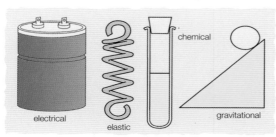

electrical

elastic

chemical

gravitational

kinetic energy

the energy possessed by an object as a result of its movement (including heat energy). equal to half the mass multiplied by the square of the velocity. Since energy cannot be destroyed. this is equal to both the energy needed to accelerate the object to its current velocity. and the energy needed to stop it again.

force

the rate of application of energy to change the velocity or shape of an object. Forces always exist in equal and opposite action-reaction pairs—a football hitting a wall has a force exerted on it by the wall that brings it to a halt. but it exerts an equal force on the wall (and thus Earth). Since the football's mass is far smaller than Earth's. the change in the Earth's velocity is around 15.000.000.000.000.000.000.000 times smaller than the velocity the ball had before hitting the wall. The total change (equal to the force multiplied by the time it was applied for) is sometimes called the impulse.

newton (N)

the SI unit of force. A newton is the amount of force needed to accelerate a mass of 1 kg by 1 m s^{-2}.

dyne (dyn)

an outdated small unit of force. not in general use. A dyne is the force required to accelerate a mass of 1 gram by 1 cm per second per second. One hundred-thousand dyne = 1 newton.

thrust from propellers
on boats and planes

thrust from swimming

rocket thrust

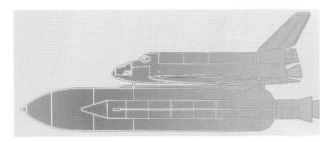

*Propellers, wheels and
people all generate thrust
by pushing directly against
their surroundings. A rocket's
thrust is generated by creating
a large amount of hot gas then
accelerating that in the
opposite direction to the
rocket's movement.*

❧ pound-force (lb-f)

the force exerted by the Earth's gravity on a 1-lb weight, equal to just under 32 lb.ft.s^{-2} (**pound feet** per second squared) or about 4.4 newtons.

thrust

the creation of a force by pushing material backwards relative to the desired direction of movement, leading to forward movement as a result of the equal and opposite reaction force. Examples of acceleration from thrust include propellers on planes or boats, rockets, and jumping or swimming. The greater the ratio of the thrust to the object's weight, the greater the acceleration produced. A rocket or other object intended to travel straight upward must produce a thrust greater than its weight, or it will not be able to lift off the ground. Thrust is usually measured in either pounds or newtons. The thrust produced is equal to the mass of the material pushed backward multiplied by the acceleration given to it.

dynamometer

a machine to measure the power (and torque) produced by an engine (of any kind), or the power input required by a piece of machinery.

❧ gearing

the use of gears (or belts and pulleys) to transfer force from one place to another, usually changing the speed of rotation (and the

*A pair of meshed gears can
be used either to increase
torque (lowering speed) or
increase speed of rotation
(lowering torque).*

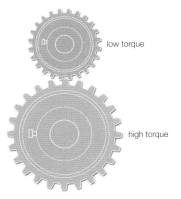

low torque

high torque

torque) in the process. Note that a gearing system cannot change the amount of power (which is always slightly less coming out than going in, as a result of frictional losses), but can increase torque by decreasing speed of rotation or vice versa.

ஜ moment

a measure of the turning effect caused by a force around a pivot, equal to the force multiplied by the distance from the pivot. Thus the rotational effect of a 1 newton force 2 meters from the pivot is equal to that of a 2 newton force 1 meter from the pivot. If two such forces try to rotate the object in opposite directions, they cancel each other out, and the rotational speed does not change. Moment has the same dimensions as energy, but is measured in different units (for the SI system, newton-meters). A moment of 1 newton-meter uses 1 joule of energy to move through an angle of 1 radian. Torque is another word for moment, applied particularly

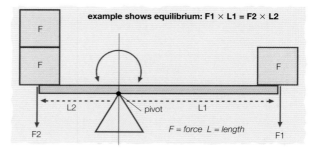

Equal and opposite moments generated by non-equal forces cancel out around a pivot.

to engines and motors. Moment is the vector product of the force vector and the vector connectingh the pivot to the point at which the force is applied.

pound foot (lb ft)

a unit of torque or moment, being the torque produced by a force of 1 pound-force applied 1 foot from the pivot point.

power

the amount of energy used (or work done) per unit time.

watt (W)

the SI unit of power, equal to 1 joule per second (or, in electrical units, a current of 1 ampere across a **potential difference** of 1 volt). For many purposes, power consumption of modern equipment is often stated in kilowatts, and commercial power generation rarely deals with quantities smaller than a megawatt.

efficiency

in energy (and heat transfer) terms, the efficiency of a machine or process is the proportion of the energy put in that is used to do useful work, stated either as a fraction or a percentage. A notional perfectly efficient machine (with an efficiency of one) would require no energy to function, and waste no energy to its surroundings as unwanted heat—but such a machine is impossible in practice. Internal combustion engines tend to have an efficiency of below 20 percent, while steam turbines in power stations are around 35 percent efficient.

horsepower (hp)

a very old unit of power, originally defined by the scientist James Watt as the average rate of work of a typical horse walking on a treadmill to power machinery. Later defined numerically as 550 **foot pounds** per second (746 W). Often used as the measure of power for internal combustion engines.

brake horse power (bhp)

the maximum available output power of an engine (measured in horsepower). The term is still used in the U.K., but fell into disuse in the U.S. after some car manufacturers produced unrealistically high figures by removing vital parts of the engine. Both the modern U.K. use of bhp and the American "hp (**S.A.E.**)" are based on complete production engines.

work

the energy transferred between objects by applying a force over a distance, equal to the **scalar** product of the force and movement (both of which are **vectors**). If the movement is directly along the line in which the force is applied, this is simply the magnitude of the force multiplied by the distance moved. Work has the same units as (and is essentially a special case of) energy.

foot pound (ft lb)

an old unit of work (or energy), defined as the work done by a pound-force acting over a distance of 1 foot, equal to about 1.36 J. Although it has the same units as the **pound foot**, the two should not be confused—the difference is that with a foot pound, the force moves along the foot in the unit, whereas with a pound foot the measured foot is at right angles to the direction of the force.

erg

a little-used unit of work, far too small to be of practical use, defined as the work done by a force of 1 dyne acting over a distance of 1 cm. One joule is equal to 10 million ergs; 1 horsepower is 7.5 billion ergs per second.

ergometer

occasionally used to mean a **dynamometer**; more often a medical device for measuring the work a muscle is capable of doing. Usually a stationary rowing machine for practising on dry land, rather than either of these, however.

speed and flow

speed

the rate of an object's motion. equal to distance traveled per unit time. Speed is a **scalar** quantity: 30 mph in any direction is 30 mph.

velocity

the rate at which an object's location changes with time. It is the **vector** equivalent of speed. and measures both the rate and the direction of the object's movement. The magnitude of an object's velocity is its speed.

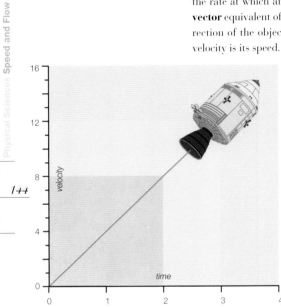

⟡ acceleration

the rate at which an object's **velocity** or **speed** changes with time. It is therefore measured in change of speed per unit time—i.e.. distance per unit time per unit time (in SI units. meters per second squared: ms⁻²). For velocity. a **vector** acceleration is required. while change of speed is **scalar**. Negative acceleration in terms of speed is often called deceleration but the term is not normally used with changes in velocity. Where a vector acceleration is not exactly parallel to the original velocity. the direction of movement changes. usually as well as the speed.

A geostationary satellite in circular motion is an example of continuous acceleration while maintaining a constant speed.

meters per second (m s⁻¹)

the SI unit of **speed** (and **velocity**). One m s⁻¹ is equal to 3.6 km/h. or a little over 2 mph. Used for most scientific work that involves speed or velocity.

miles per hour (mph)

the normal measure of a vehicle's speed primarily in the U.S. and U.K. One mile per hour is about 0.45 meters per second.

⟡ Doppler effect (Doppler shift)

the apparent change in the frequency (and wavelength) of a wave caused by the movement of the observer relative to the source of the waves. The effect is the same for a given relative velocity even if either the source or observer is stationary. For example. to a listener standing still on the ground. as a plane flies overhead the sound of the engine appears to change as its velocity changes. The engine's apparent frequency increases as the plane approaches and decreases as it flies away. The phenomenon was first proposed by

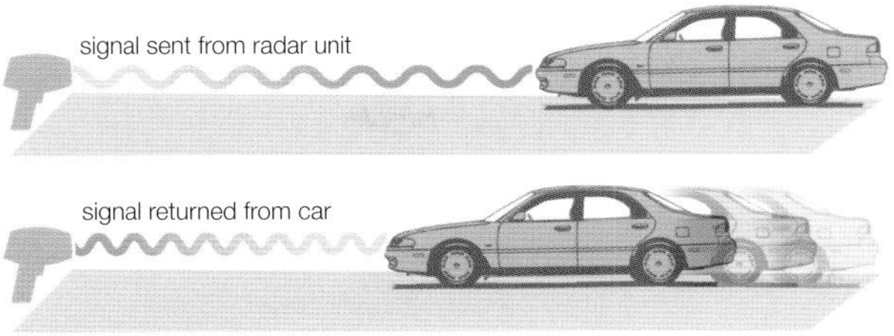

signal sent from radar unit
signal returned from car

The faster a car moves toward you, the more the Doppler effect compresses the sound waves and the higher the pitch of the sound you hear.

Austrian mathematician Christian Doppler (1803–53) to explain the **redshift** in astronomy—which is now thought to be caused by a similar but different effect.

kilometers per hour (km/h)

a measure of speed used for vehicles in most countries, and equal to about $^2/_3$ of a mile per hour. Speeds in km/h are occasionally quoted as "kph," but since the SI symbol for kilometers is km (not k) this is technically incorrect.

feet per second (fps; ft s^{-1})

a measure of speed in scientific experiments, now little used. One fps is about equal to 0.3 meters per second, or 1.1 km/h.

knots

the unit of speed for ships (and one of those used for aeroplanes). One knot is 1 nautical mile per hour (about 1.15 "land" miles per hour, or 0.51 meters per second). The **air speed** of an aircraft is usually measured in knots, while the ground speed is usually recorded in either kilometers or miles per hour. Occasionally (and informally) also used to mean a distance of 1 nautical mile.

❧ angular velocity

the velocity of rotation of a spinning object. Since it is necessary to specify a direction (the axis around which the object is spinning, the poles of which remain stationary) this unit is a **vector**. It can also be used to describe the speed at which one object is orbiting another, although this becomes complicated if the object being orbited is also rotating. Angular velocity is measured in angle per unit time; the SI unit is radians per second. Rotational speed is similar but different, being the number of complete rotations made per unit time.

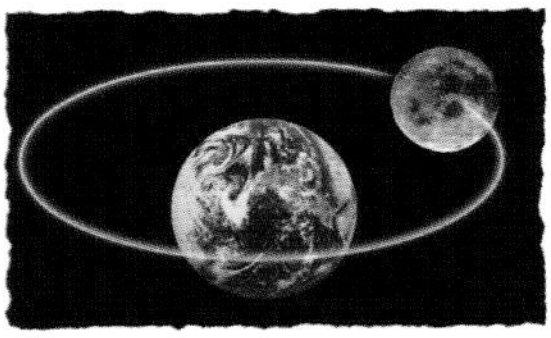

Appearances can be deceptive—the moon orbits Earth once a month, but the Earth's rotation means that the moon is only visible from one place for a few hours—rather than days—at a time.

revolutions per minute (R.P.M.)

a unit of rotational speed. A rotational speed of one R.P.M. is equal to an **angular velocity** of 6° per second. Objects with speeds measured in R.P.M. include C.D.s (100s), car engines (1,000s) and gas turbines (up to tens of thousands). Note, however, that the term is also used as an abbreviation of "rounds per minute," the rate at which a weapon fires bullets or other projectiles.

tachometry

the measurement of angular or rotational speeds. Any instrument that records a **speed** or **velocity** of rotation is a tachometer. Most often seen as a dial on a car dashboard that tells you how fast the engine is spinning, a tachometer can serve many other purposes, e.g., a car's speedometer works by measuring the speed of rotation of the tires and multiplying by the circumference of the tire. More generally, a wide variety of industrial and power-generation machinery rotates, often at high speed, and a change in rotational speed can be an indication of problems that must be checked.

❧ momentum

a measure of how difficult it is to stop a moving object. Momentum (equal to **mass** times **velocity**) is not the same as **kinetic energy** (half of mass multiplied by the square of speed)—a 10 kg rocket moving at 1,000 mph and a 1 tonne car moving at 10 mph have identical momentum but the kinetic energy of the rocket is a hundred times higher. In addition to the difference in magnitude, momentum is a **vector**, while kinetic energy is **scalar**. So if two identical objects collide while traveling in opposite directions at the same speed and both stop, their combined momentum (which was zero before they hit) remains the same, but their kinetic energy changes drastically, all being converted to other forms of energy, such as noise and the work done deforming car bumpers.

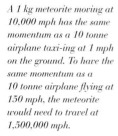

A 1 kg meteorite moving at 10,000 mph has the same momentum as a 10 tonne airplane taxi-ing at 1 mph on the ground. To have the same momentum as a 10 tonne airplane flying at 150 mph, the meteorite would need to travel at 1,500,000 mph.

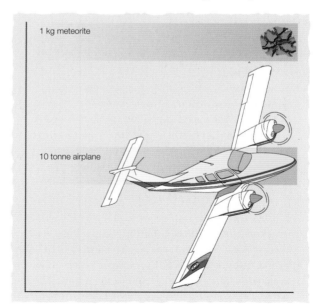

1 kg meteorite

10 tonne airplane

muzzle velocity

the speed of a bullet or other projectile as it leaves the barrel of a gun or other launcher. Since there is no directional information (the bullet is always moving along the length of the barrel unless something is badly wrong), this is strictly a speed, not a velocity.

Mach number

the speed of an object relative to its surroundings (e.g., the **air speed** of an aircraft) expressed as a multiple of the speed of sound. An object with a Mach number greater than 1 is termed supersonic; when the Mach number reaches 5 it becomes hypersonic. Since the speed of sound varies with the density of the air through which it is traveling, Mach 2 at sea level (1,520 mph) is not the same as Mach 2 at a typical cruising altitude of 50,000 feet (1,320 mph).

air speed

the speed of an aircraft or other vehicle relative to the surrounding air. Since the air, especially at the altitudes used by modern jet planes, is usually moving relative to the ground, air speed is almost never the same as the vehicle's ground speed. For long-distance airliners, the difference can be significant, since the aircraft's speed is generated by thrust against the air, its maximum speed is an air speed not a ground speed. An aircraft with a cruising speed of 500 mph and a 50 mph tailwind will have an effective ground speed of 550 mph, and arrive at its destination early.

ground speed

the speed of a vehicle, usually a plane or ship, relative to the ground. Ships also have a water (or sea) speed, which is equivalent to an plane's **air speed**.

sinking speed

the vertical component of a descending object's velocity—that is, the rate at which it is moving downward. Most often used of aircraft (especially gliders), but may also be applied to any other object in a gas or liquid (e.g., submarines or divers).

terminal velocity

the speed at which the air resistance to downward movement of a falling object is exactly equal to the gravitational force acting on it. If the object is traveling at a speed below terminal velocity, it will speed up until it reaches terminal velocity. If it is traveling at a speed above terminal velocity it will slow down until it reaches terminal velocity. If the falling object's speed is not equal to its terminal velocity, the speed will get close to the terminal velocity. Terminal velocity of a human being (with parachute closed) ranges from about 120 mph with limbs spread to over 200 mph in a streamlined "diving" position.

In theory, people on two airplanes traveling in opposite directions will see time pass at a different speed on the other airplane to the passage of time on their own. In practice, such effects are only significant at very high fractions of the speed of light.

speed of light (c)

the speed of light in a vacuum, exactly 299,792,458 meters per second (or approximately 670 million mph), which the laws of physics say can never be exceeded. Any observer will measure the value of c relative to themselves as being the same, however, regardless of their **velocity** relative to the source of the light. This leads to a variety of extraordinary effects, of which the apparent distortion of time and distance experienced by a person or object traveling at speeds close to c is probably the best known. The exact speed given in the first sentence is defined, not measured: the modern definition of a meter is the distance light travels in $1/299,792,458$ of a second.

critical velocity

the minimum speed relative to its surroundings at which a liquid or gas will display turbulence. Below this speed, the flow is "smooth," and turbulence is prevented by the fluid's **viscosity**.

viscosity (dynamic viscosity)

a measure of the resistance of a fluid (a liquid or gas) to flow, both internally and through what may be called "fluid friction." (Although solids can flow, albeit very slowly, this is an entirely different process—see **creep**—and solids do not strictly have a viscosity.) Viscosity is independent of pressure, but does change with temperature—as the temperature increases, the viscosity of a gas will rise, while that of a liquid tends to fall.

kinematic viscosity

the dynamic **viscosity** of a fluid divided by its density. This can be a more useful measure of a fluid's behavior than dynamic viscosity alone, especially where flow under gravity is involved. If two fluids (e.g., honey and machine oil) have the same dynamic viscosity but different densities, they will behave differently—and their kinematic viscosity reflects this difference.

Poise (P)

an old but still useful unit of dynamic **viscosity**, defined as one dyne second per square centimeter (0.1 Pa S), the viscosity of water at room temperature is around one centipoise. The equivalent unit of kinematic viscosity is the Stokes (St), defined as 1 Poise per gram per cubic centimeter. This unit is named after the scientist Sir George Stokes, so the singular is, unusually, 1 Stokes. There is no special name for the SI equivalent of either unit (Pa S for dynamic viscosity and $m^2 s^{-1}$ for kinematic viscosity).

Rhe

a unit of fluidity (the inverse of dynamic **viscosity**). The fluidity of a substance in Rhes is one divided by its viscosity in Poise, giving the unit a value of 10 per Pascal second. It was originally defined in terms of the centipoise (1,000 per Pa S), and is occasionally still used this way.

viscosity grade (V.G.)

one of several measures of viscosity, usually used for commercial lubricants and oils. **S.A.E.** viscosity grades for engine oil were originally a measure of the length of time a fixed quantity of oil took to flow through a test orifice at 100°C, but have been re-defined in terms of Stokes. I.S.O. (and A.S.A.) viscosity grades are equal to the kinematic viscosity of the fluid in Stokes at 40°C, with a maximum deviation specified for each grade. The viscosity index is a measure of how much the viscosity of a fluid changes with temperature.

rheometer

any machine for measuring deformation or flow, whether in a solid, liquid or gas. A viscometer, for measuring the viscosity of a fluid, usually works either by recording the time taken for the liquid to flow along a tube, or by dropping a metal ball in and measuring the speed at which it falls. Other types of rheometer are used with a wide range of materials, applying force in a variety of ways (e.g., heat, rotation, squeezing, cutting—or a combination of two or more of these) and recording the behavior of the substance being tested.

❧ water inch

a rather poorly defined unit of flow, being the amount of water per unit time that will pass through a round hole one inch in diameter. Even with the restriction that gravity is the only force applied, the rate of flow still depends on the depth of water above the hole. If the water is pouring from the bottom of a tank 330 feet (100 m) high, it will flow much faster at high pressure when the tank is full than at low pressure when it is empty.

cusec

a unit of flow rate: 1 cubic foot per second. The metric equivalent is a cumec, equivalent to 1 cubic meter per second (just over 37 cusec).

gallons per minute (gal min⁻¹)

a small unit of flow rate: 1 U.S. gallon per minute is only 0.0022 cusec, making 1 cusec equal to 448 gallons per minute.

megaliters per hour (ML/hr)

a measurement of flow rate used in industrial processes, water supply and the study of rivers. It equals one million litres per hour, or about $10^{1/4}$ cubic feet per second. For major rivers, or very large-scale industrial measurements, flow will sometimes be specified in megaliters per minute, or even per second.

empty tank: low pressure

full tank: high pressure

The higher the pressure, the faster water flows through a standard-sized hole.

mass and weight

Because of the hugely different sizes and densities of the planets of the solar system, their gravitational forces are correspondingly different. So, a mass of 1 kg would feel like only 166 grams on the moon, and 2.364 kg on Jupiter.

1   *Jupiter 2.364*
2   *Mercury 0.378*
3   *Saturn 1.064*
4   *Venus 0.907*
5   *Uranus 0.88*
6   *Earth 1*
7   *Neptune 1.125*
8   *Moon 0.166*
9   *Pluto 0.067*
10   *Mars 0.377*

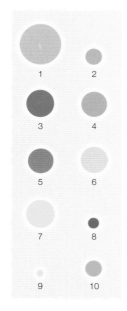

‡ weight

the gravitational force acting on a body, measured in **newtons** or **dynes** in the SI system, or **pound-force** in the U.K. imperial system. In many scientific contexts, it is essential to distinguish between weight and mass, and strictly speaking the base units of the metric and imperial systems (kilograms and pounds) are units of mass: it is, however, quite usual to come across these units loosely used to mean the mass under the force of the Earth's gravity at its surface—in fact the imperial system defines pounds as weight under this assumption. The difference becomes immediately apparent when comparing the weight of a body on Earth, and then on the moon; although its mass is unchanged, its weight is considerably less due to the moon's weaker gravitational force. But for most practical purposes, on this planet at least, the terms weight and mass are virtually synonymous.

‡ mass

the amount of matter in a body, the inertial mass being a measure of its resistance to change of motion, and the gravitational mass a measure of its attraction to another body. The base units of mass in use today are mainly the kilogram in the SI system, and the pound in the U.S. Customary and U.K. imperial systems.

‡ gravity

the force of attraction between bodies due to their mass. This force was first defined by Isaac Newton (1642–1727) in his Law of Gravitation, which states: "Any two particles of matter attract one another with a force directly proportional to the product of their masses and inversely proportional to the distance between them." This is expressed in the equation, $F = G \, (m_1 m_2 / d^2)$, where F is the gravitational attraction between bodies m_1 and m_2, and d is the distance between them; G is the gravitational constant, equal to 6.6732×10^{-11} n m^2 kg^{-2}.

g

the symbol for acceleration due to gravity of a body in free fall. It is also used loosely as a unit of acceleration, where 1 g is equal to 9.80665 meters per second squared (about 32.17405 feet per second per second) at the Earth's surface, but this varies in practice with altitude and latitude. The g is used when comparing the gravitational forces of different bodies such as planets, and also when measuring the force exerted on an accelerating (or decelerating) body, especially in aeronautics and in space travel.

density (ρ)

the ratio of the **mass** of a body to its volume. Density is expressed in terms of mass per unit volume, such as kilograms per cubic meter, or pounds per cubic foot. Relative density, also known as specific gravity, is the ratio of the mass of a given volume of material to the mass of the same volume of water at 4°C (i.e., at maximum density).

center of gravity

the point where the resultant of gravitational forces experienced by all the particles of a body acts. In a uniform gravitational field such as at the Earth's surface, it is the same as the center of **mass**: the point where the mass of the body may be considered to be concentrated, where the resultant of external forces may be considered to be acting, and the point through which any plane would divide the body into two parts of equal mass.

	gravity variation				
latitude (°)	altitude (m) 1000	10000	20000	30000	
0	9.7805	9.7774	9.7496	9.7188	9.6879
5	9.7809	9.7778	9.75	9.7192	9.6883
10	9.782	9.779	9.7512	9.7203	9.6895
15	9.784	9.7809	9.7531	9.7222	9.6914
20	9.7865	9.7835	9.7557	9.7248	9.694
25	9.7897	9.7866	9.7589	9.728	9.6971
30	9.7934	9.7903	9.7626	9.7317	9.7008
35	9.7975	9.7944	9.7666	9.7358	9.7049
40	9.8018	9.7987	9.7709	9.7401	9.7092
45	9.8063	9.8032	9.7754	9.7446	9.7137
50	9.8108	9.8077	9.7799	9.7491	9.7182
55	9.8151	9.812	9.7834	9.7534	9.7226
60	9.8192	9.8161	9.7884	9.7575	9.7266
65	9.823	9.8199	9.7921	9.7613	9.7304
70	9.8262	9.8231	9.7953	9.7644	9.7336
75	9.8287	9.8257	9.7979	9.767	9.7362
80	9.8307	9.8276	9.7998	9.7689	9.7381
85	9.8318	9.8287	9.801	9.7701	9.7392
90	9.8322	9.8291	9.8013	9.7705	9.7396

Note: internationally agreed standard gravity on Earth = 9.80665 meters per second per second

As Earth is not a perfect sphere, and because gravitational force is dependent on the distance from the center of Earth, there is appreciable difference in the force of gravity according to both latitude and altitude.

troy weight system

a system of weights used for measuring precious metals and gemstones. The troy system was in use in Britain until the 19th century, but has been superseded first by the avoirdupois system, and more recently by metric units, although it survives to some extent in the jewelry trade in the U.K. and North America. As with many traditional units, its usage varies slightly between the U.K. and U.S.

avoirdupois weight system

the system of weights in use in Britain from the 14th century to the introduction of metrication in the 1960s, and widely used in the English-speaking world up to the present day. The base unit of the system is the **pound** (lb). The term derives from the old French *avoir du pois* (meaning "having weight").

apothecaries' weight system

a system of weights originally used by apothecaries for measuring very small amounts. First used in the 17th century in Britain in the pharmaceutical trade, it was based on the older troy ounce, and was eventually superseded by the troy weight system, which itself is now being replaced by metric measures. It differs from the troy

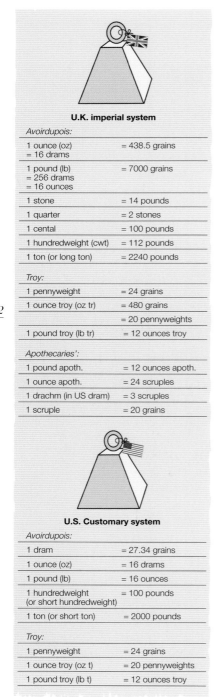

U.K. imperial system

Avoirdupois:

1 ounce (oz) = 16 drams	= 438.5 grains
1 pound (lb) = 256 drams = 16 ounces	= 7000 grains
1 stone	= 14 pounds
1 quarter	= 2 stones
1 cental	= 100 pounds
1 hundredweight (cwt)	= 112 pounds
1 ton (or long ton)	= 2240 pounds

Troy:

1 pennyweight	= 24 grains
1 ounce troy (oz tr)	= 480 grains
	= 20 pennyweights
1 pound troy (lb tr)	= 12 ounces troy

Apothecaries':

1 pound apoth.	= 12 ounces apoth.
1 ounce apoth.	= 24 scruples
1 drachm (in US dram)	= 3 scruples
1 scruple	= 20 grains

U.S. Customary system

Avoirdupois:

1 dram	= 27.34 grains
1 ounce (oz)	= 16 drams
1 pound (lb)	= 16 ounces
1 hundredweight (or short hundredweight)	= 100 pounds
1 ton (or short ton)	= 2000 pounds

Troy:

1 pennyweight	= 24 grains
1 ounce troy (oz t)	= 20 pennyweights
1 pound troy (lb t)	= 12 ounces troy

weight system in its subdivisions of the troy ounce, known in this system as the apothecaries' ounce. The apothecaries' ounce is divided into **drams** (or drachms), **scruples** and **grains**.

tonne (t)

an SI unit of mass, equal to 1,000 kilograms. To distinguish it from the U.K. imperial **ton** (to which it is very close in value) and the U.S. ton, it is also sometimes called a metric tonne or metric ton. The word tonne is closely related to the tun of liquid measurement, which is, confusingly, sometimes also called a tonne.

ton

a unit of mass or weight in the U.K. imperial and the U.S. Customary systems, but with different values in each system. The ton in common usage in the U.K., also known as the long ton, is equal to 2,240 lb (1,016.0416 kg, or 1.0160416 tonne), whereas the U.S. ton, or short ton, is equal to 2,000 lb (907.18474 kg or 0.90718474 tonne). Historically, the ton was also used as a unit of volume, especially in shipping dry goods, and, until metrication, variations on the term often referred to units of volume, such as the register ton, shipping ton, and hoppus ton (*see* **hoppus foot**).

hundredweight (cwt)

a unit of mass or weight in the U.K. imperial and U.S. Customary systems, but, like the ton, of which it is a subdivision, with different values in each system. In its abbreviation, cwt, the c is the Roman symbol for one hundred. In Britain and many other English-speaking countries, it is equal to 112 lb, or $1/20$ of a (long) ton (50.80208 kg), but recent usage in North America has produced a unit of 100 lb, known as a short hundredweight, $1/20$ of a short ton (45.359237 kg). This latter is exactly equivalent to the less commonly used unit known as a cental in the U.K.

slug

a unit of mass in the gravitational system. It represents the mass which, when acted on by a force of 1 pound force (1 lbf) has an acceleration of 1 foot per second squared (1 ft s^{-2}), and is equal to about 32.174 lb (14.594 kg). A metric technical unit of mass known as the glug, or metric slug, is now more widely accepted than the slug (though not yet recognized as an official SI unit). It represents the mass which, when acted on by a force of 1 kilogram-force (1 kgf) has an acceleration of 1 meter per second squared (1 kg s^{-2}).

stone

a mainly British unit of weight, equal to 14 lb. Nowadays its use is restricted almost entirely to descriptions of body weight, but even there is disappearing in favor of the kilogram. The name stone is thought to derive from the use of actual stones or rocks of specific weight used on balance scales.

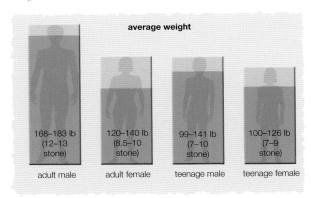

Despite metrication, body weight is still normally expressed in stone and pounds in the U.K.

kilogram (kg)

the base unit of mass of the SI system, one of the seven base units from which all other units are derived. It is specifically a unit of mass, rather than weight or force, unlike the base units of traditional systems where these were effectively interchangeable. The International Prototype Kilogram, a cylinder of platinum-iridium alloy, is the definitive standard for the kilogram internationally, and is kept by the Bureau International des Poids et Mesures in Sèvres, France. Roughly speaking, the kilogram has the mass of 1 cubic decimeter (or 1 liter) of water at 4°C.

pound (lb)

the base unit of mass in the U.K. imperial and U.S. Customary systems, equivalent to 0.45359237 kg. The use of a pound in one form or another dates back to ancient Rome, where the *libra pondo* (pound of weight) was commonly used, and from which derive many traditional European units of weight; it is also the source of the abbreviation lb, short for *libra*. The U.K. and U.S. **avoirdupois** pound is divided into 16 ounces, unlike the Roman **libra** and similar units used in southern Europe which divided into 12, though the old practice of division by 12 survived in the English troy system. It is sometimes important to distinguish between the avoirdupois pound. which is still in use in the U.S. and informally in the U.K.. and the now obsolete **troy** pound or the rarely used **apothecaries'** pound. when the abbreviations lb av. lb t. and lb ap are used. Troy and apothecaries' pounds are identical in value. and equal $^{144}/_{175}$ (0.822858) avoirdupois pounds. or 0.373.242 kg. An additional distinction is made by the abbreviation lbf. for **pound force**. which is a unit of force rather than mass: 1 pound force is equal to 4.448221615 newtons.

libra

a unit of weight in ancient Rome. still in use informally in many Mediterranean and Spanish-speaking countries. Like the name of the constellation and the horoscope sign. the Latin word *libra* originally referred to the balance scales used for measuring the weight. The unit of weight had a variable value. but came to refer specifically to around 0.722 U.K. pounds (0.32745 kg). which was divided into 12 *unciae*. The libra has survived throughout Europe and its colonies. with various values very approximately the same as the U.K. pound. even where the SI system has been officially adopted. In France it is known as the *livre*. and has come to mean exactly 500 grams. and in Italy the *libbra* has varied in size from the ancient Roman value to around one U.K. pound—though confusingly it is used today as a synonym for the kilogram.

The shekel was a common unit of weight throughout the ancient Middle East. but is probably better known today as the coin of the same weight used by the Hebrews.

๛ shekel

an ancient Babylonian unit of weight used widely in the Middle East. and adopted by the ancient Hebrews. Its exact size is disputed. but values range from around 8 to 16 grams. Some sources confidently give an exact equivalent of 252 grains (about 16.33 grams). others give 8.4 grams: it is probably safer to assume that we do not know. The shekel was also a Hebrew coin of the same weight.

dram (drachm; dr)

a unit of weight in the **avoirdupois** and **apothecaries'** systems. in which it has different values: one avoirdupois dram (1 dr av) is equal to $1/16$ of an ounce (about 1.7718 grams), whereas one apothecaries' dram (1 dr ap) is equal to $1/8$ of a troy ounce (about 3.8879 grams). In the U.K. the spelling "drachm" is common (presumably to distinguish it from the **dram** still in informal use as a measure for whisky and other spirits), and this confirms its derivation from the Greek *drachme* ("a handful").

penny-weight

a unit of weight in the troy system, equal to 24 grains, or $1/20$ of a troy ounce (1.552 grams).

scruple

a traditional unit of weight in the apothecaries' system, equal to 20 grains or $1/24$ of an apothecaries' ounce (about 1.296 grams). The scruple has been in use in Europe since Roman times, with a similar value to the modern scruple, and the word derives from the Latin *scrupulus* (a small sharp stone).

gram (g)

a small unit of mass in the SI system, at one time the base unit of mass of the metric system known as the C.G.S. (centimeter-gram-second) system, but now defined as $1/1000$ of the International Prototype Kilogram. The unit derives from the Greek *gramma*, which was similar in value to the Roman scruple. The original French spelling *gramme* is still seen occasionally, but is not internationally recognized.

grain

a small unit of weight in the U.K. imperial and U.S. Customary systems. Its name derives from an early definition as equivalent to the weight of one grain of wheat or one barleycorn. It was in effect the original base weight of the British systems, as it was used to define the pound in each usage: one **avoirdupois** pound = 7,000 grains, one **troy** pound = 5760 grains. The term grain is also used in the jewelery trade as equal to one-quarter of a **karat** (50 milligrams); in this context it is sometimes called the pearl grain after its use for measuring the weight of pearls.

point

a small unit of mass used for measuring gemstones, equal to one-tenth of a **karat** (20 milligrams).

Planck mass

a very small unit of mass, the smallest possible mass according to the laws of physics as they stand today. Like the units of Planck length and Planck time, it is derived from the **Planck**, and is named after Max Planck (1858–1947), who developed the quantum theorem. The Planck mass unit corresponding to units of Planck length and Planck time is equal to 2.1×10^{-8} kg.

Technology
and Leisure

computers and communications

kibi-

part of an attempt to solve ongoing arguments about the use of existing standard prefixes in computer terminology. The problem occurs because some computer components (e.g., memory) expand more logically by powers of two than powers of 10. Thus, a gigabyte of memory is 1,073,741,824 (3^{30}) bytes. Hard drives, however, have no physical reason to follow this pattern, and manufacturers therefore quote sizes using the standard decimal gigabyte of 1,000,000,000 (10^9) bytes. Hardware speeds (including bandwidth) are always quoted in decimal versions, as are D.V.D. sizes, while C.D.s use binary units. The International Electrotechnical Commission (I.E.C.) suggested a set of new prefixes defined in powers of 2 (kibi- = 2^{10}, mebi- = 2^{20}, gibi- = 2^{30}, etc.) to allow distinction between the two sizes, but these are not in wide use. To make matters worse, some pieces of hardware have unique definitions—in a 1.44 M.B. floppy disc, the M.B. is (1000 times 1024) bytes, and the "k" in a 56 k modem is approximately 1,028.6 bits (the nominal speed is 57,600 baud), among many others.

✷ bit (b)

the smallest unit of computer memory, a bit is a single one or zero. Most modern computers deal with bits in larger groups, but "embedded" high-tech systems with smaller memories (e.g., mobile phones, factory robots) often use individual bits to keep track of simple on/off conditions (whether a phone is in silent mode, for example). For larger numbers, different systems can be either big-endian (starting with the most important digit, like normal decimal numbers) or little-endian (starting with the smallest digit).

binary number	decimal equivalent (big-endian)	decimal equivalent (little-endian)
1000	8	1
0001	1	8
00000001	1	128
10010001	145	137
1110	14	7
111110	62	31
011111	31	62
1111	15	15

Big-endian (starting with the largest digit) and little-endian (starting with the smallest digit) numbers both have their uses, but will use the same arrangement of bits to indicate very different numbers.

byte (B)

a byte consists of eight bits, recorded (and acted on) as a group. It therefore has 256 (2^8) possible values, traditionally either from zero to 255, or −127 to +128.

nybble (nibble)

a nybble is half a byte, four bits giving it 16 possible values. Most programming languages have no method for accessing memory in nibbles, but where memory is in short supply nybbles can be used to record a variable condition that does not require a full byte, such as ring-tone volume or position in a complex process.

word

the unit of memory dealt with internally by a computer's central processing unit (C.P.U.), and passed between C.P.U. and the main memory. The size of either or both can vary between different types of computer, though most modern PCs use a 32-bit word. However, embedded systems often still only use 16-bit (or even 8-bit) words, while high-end computers may have a 64-bit (or larger) word.

✿ floating point

The table shows how floating point works using a 3-digit decimal number (i.e., from –999 to +999) and a power of ten in the range +3 to –3. As you can see, there is no way to store 4,002 (or 4.002).

a floating point number is actually two numbers stored together, each occupying a fixed number of bits specified by the computer and programming language used (though most modern computers follow the I.E.E.E.'s 754 standard). These two numbers represent an integer and a power of two by which that integer is multiplied to give the floating point value. While this system allows a very wide range of numbers to be stored in a single relatively compact format, it results in the number stored being only an approximation. This can cause problems when adding small numbers to large ones, or where the last few digits of a number are the important part, as a floating point system may be unable to distinguish between 1.000.000.000 and 1.000.000.004.

integer	power of 10	no. stored
42	0	42.0
42	+2	4200
42	-2	0.42
426	-2	4.26
-426	-2	-4.26
426	+1	4260
427	+1	4270
-427	+1	-4270
427	+3	427000
427	-3	0.427
4	-3	0.004
4	+3	4000

character

a single letter, digit, punctuation mark or similar. Typically represented in modern computer systems using a full byte, though ASCII character codes require only seven bits (so are sometimes used for transmitting large amounts of text). Early telegraph systems often used only five or six bits, to help conserve their limited bandwidth.

string

a sequence of characters "strung" together: the usual way of storing text in a computer program. The term is sometimes also used (with an appropriate modifier word) for connected sequences of other memory units, e.g., a "binary string" is a string of bits.

resolution

the amount of detail present in a picture or displayed by a screen. The higher the resolution, the better the image will tend to look, and the more it can be magnified before appearing blocky. A computer monitor has a "native" resolution, as the display surface is composed of a large number of pixels, but the monitor will also be able to display other resolutions if necessary. However, these other resolutions may either appear blurred or have jagged edges (especially on LCD monitors), as pixels cannot be assigned partially to one colour and partially to another.

D.P.I. (Dots Per Inch)

the standard measure of resolution for scanners and printed images. Exactly as you expect, the number of spots along a 1-inch line at which the scanner records the color, or the printer places a dot of ink. Since this is a linear measure and images cover an area, doubling the resolution in D.P.I. will quadruple the size of a scanned image file (or reduce the printed image to quarter-size).

pixel

the unit of display in the memory of a computer (or other digital system) or on a video screen—the name is a contraction of "picture element." Very early computers, where the only control over the pixel was whether it was on or off, were followed by systems with higher resolutions (320×200 pixels was fairly typical in the early 1980s) and multiple (often 8 or 16) colors. Modern systems usually have a screen resolution of 1024×768 or higher, and the ability to display millions of different colors—and even cheap digital cameras can capture images with millions of pixels.

voxel

the 3D equivalent of a pixel, widely used for representation of scientific data (e.g., medical imaging, such as M.R.I. scans) and, less often, for certain types of computer game. A genuinely 3D display would also have a resolution measured in voxels.

rrrrrr

1 8 d.p.i.
2 16 d.p.i.
3 32 d.p.i.
4 64 d.p.i.
5 128 d.p.i.
6 256 d.p.i.

The resolution used increases from left to right, with each step to the right making the letter much crisper, hence easier to read.

159

The storage of 3D information as voxels allows medical imagers to examine any cross-section through it at any angle.

benchmark/specmark

a standard test (or series of tests) designed to measure the speed of a computing system (or one of its components). Benchmarking allows comparisons to be made between different items designed to perform the same task. S.P.E.C. (the Standards Performance Evaluation Corporation) benchmarks are designed carefully to simulate real usage.

dhrystone

a benchmark speed test for personal computers, using no floating point numbers and designed to test both the power of the computer and the efficiency of the "compiler" that converts the standard program, written in a language called "C," into code for the specific machine. Named as a pun on an older benchmark called the Whetstone, it is now falling into disuse as tests more representative of real use become widespread.

MIPS (Millions of Instructions Per Second)

a badly-flawed measure of computer C.P.U. speed using integers. Different processor designs use different numbers of instructions to perform the same task, and the speed will often be calculated by the manufacturer using a program carefully designed to produce a high result.

FLOPS (FLoating point Operations Per Second)

the floating point equivalent of MIPS, and only a little more useful as a genuine indicator of performance. The speed of super-computers is often quoted in teraflops.

VLF
3–30 kHz

LF
30–300 kHz

MF
300–3000 kHz

Livermore loops

a benchmark designed to measure the speed with which a computer can perform the sort of complex mathematical operations typical in simulations of nuclear physics. Named after the Lawrence Livermore Laboratory, in California.

baud

the base unit of bandwidth (speed of data transfer) for digital communications, equal to one bit per second. Low transmission speeds are

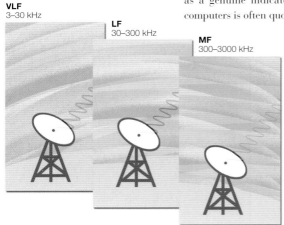

Among other things, radio antennae can detect the microwave background radiation "left over" from the start of the universe.

also sometimes quoted in bytes (or characters) per second, while higher rates (such as A.D.S.L. and cable modems) are normally given in kilobits or megabits per second, which give more impressive-looking numbers than the equivalent in bytes (1 megabit, for example, is only 125 kilobytes).

channel width

in telecommunications. the range of frequencies assigned to a single purpose. for example a single T.V. channel on a cable or satellite network. Medium wave radio usually has a channel width of 10 kHz. with 5 kHz empty bands between channels.

erlang

a unit of telephone traffic. One erlang is equivalent to one person using the line continuously (one hour of conversation per hour). If a line has enough bandwidth for 10 simultaneous conversations, and half that bandwidth is being used. the traffic is five erlangs.

attenuation

the amount by which a signal (including radio. electrical currents. laser light and earthquake shockwaves) is reduced in power as it travels from one place to another. normally measured in decibels per unit distance. Attenuation is a major problem in modern communications networks. which often need to have "repeaters" to amplify the signal at regular intervals.

frequency response

the accuracy of a system's reproduction of input at a specific frequency: theoretically this applies to any equipment. but it is generally only used for electronics (and most often for sound reproduction). Typically quoted as a range of frequencies and a variation in decibels. meaning that the object will reproduce any signal in that range to within the specified variation.

radio frequency bands

the "radio spectrum" (which includes all electro-magnetic wavelengths longer than infrared) is divided into a number of bands. each of which is allocated to specific uses.

REN (Ringer Equivalency Number)

a measure of the amount of power required to make a telephone (or fax machine. or anything else attached to a phone line) ring. This measures only the amount of power coming from the telephone line itself—a phone that plugs into the mains power supply as well can draw any amount of power from that. without affecting its REN. Typically. the maximum total REN for all the equipment plugged into a normal domestic phone line is around four or five: if the total is above this, the phones may not ring at all.

Mickey

a theoretical unit of computer mouse movement. equal to the smallest movement detectable. and named for the Disney character. Not widely used. because the actual distance varies between mice from around $1/4$ mm to 1/50 mm—and because any movement that small is likely to be caused by the natural shaking of the user's hand.

Radio spectrum divisions

VLF 3–30 kHz
radio navigation, time signals and similar simple broadcasts

LF (long wave) 30–300 kHz
aircraft navigation, some AM (amplitude modulation) broadcast radio (in Europe), amateur radio (Europe) and the U.S. "lost band" for experimental use

MF (medium wave) 300–3000 kHz
most AM radio is in this band

Medium wave 500–1700 kHz
U.S. and Canada AM radio

Short wave 3–30 MHz
a wide range of uses, including international broadcasting, amateur radio (U.S.) and "numbers" stations for communicating with spies

VHF 30–300 MHz
FM radio, T.V. stations, 2-way radio and the VOR aircraft navigation system

UHF 300–3000 MHz
mobile phone networks and TV broadcast, including HDTV

Microwave 3–300 GHz
radar, satellite communications, wireless computer networks

161

engineering

strength

the ability of a material to withstand **stress**, measured in force per unit area. Sometimes also applied to structures, in which case it is the ability of the structure to cope with load. A load is an external force applied to a structure.

❧ hardness

the resistance of a material to indentation, scratching or abrasion. It is measured by several different methods. Rockwell tests use a cone-shaped diamond (or "brale"), hard steel balls or other items to penetrate the material using a load of 10 kg, followed by a larger load of up to 150 kg, and the extent of penetration is measured. Brinell tests use a hard steel or carbide ball with a much higher load than Rockwell tests— as much as 3,000 kg—for a specific time period, which varies depending on the material. The Brinell number is load divided by the indentation area in square millimeters.

toughness

the ability of a material to withstand impact, dependent upon the distribution within the material of the stress and strain caused by the impact. The opposite of toughness is brittleness. The toughness of steel and other metals may be measured by the **Charpy impact test**.

stress

an applied force or system of forces that tends to strain or deform a body. It is defined as force per unit area, and can be expressed as kilograms per square millimeter. Stress takes a number of forms, notably shear stress (acting parallel to the surface, causing a change in shape, without particular volume change) and tensile stress (causing increase in volume or length, or both). In cases of tensile stress, the amount of deformation of a body can affect results, so the more accurate "true stress" is calculated using "instantaneous" measurements of deformation. The SI unit of stress is the **pascal**; in the U.S., the Customary (**avoirdupois**) unit pounds per square inch is often used.

strain

the amount of deformation caused by **stress**, measured as the ratio between the change in the length of an object after stress was applied and its length beforehand. When a sample is being tested for its reaction to tensile stress, it changes while the strain is being applied, so a more accurate measure, "true strain," has to be used. This involves taking "instantaneous" measurements and adding them together.

indenter

elastic recovery

permanent depth of indentation

test specimen

Using the Rockwell hardness test here, the permanent depth of indentation is the amount of penetration of the specimen after the removal of the large load, while the initial small load remains.

pascal

the SI unit of pressure or stress, equivalent to one newton per square meter (1.45×10^{-4} pounds per square inch).

standard atmospheric pressure

an arbitrary representative value for atmospheric pressure at sea level, defined as one atmosphere or 101.325 kilopascals. Atmospheric pressure tends to be measured in kilopascals, because a **pascal** is too small a unit for practical purposes. Standard atmospheric pressure was originally calculated as the pressure exerted by a column of mercury 760 mm tall at 0°C. The real value for atmospheric pressure at sea level varies depending on where in the world it is measured, the weather conditions, time of day, etc.

tension

the force exerted on an object or material by stretching it, causing it to increase in volume or length, or both. Applying tension until the substance breaks is known as tensile testing, and the reaction of the substance as it is being tested can be measured at different points, and plotted on a graph to show how it reacted at each stage.

✿ compression

the opposite of **tension**, caused by "squeezing" an object or material, and reducing its volume. For metals, results of compression tests are usually the same as those for tensile tests, but for other materials, such as polymers, this is not the case. The term compression has a particular application with regard to the internal combustion engine, and refers to the compressing of the air/fuel mixture before combustion. The compression ratio is the ratio between the largest volume in a cylinder and the smallest, when the piston has reached maximum compression.

elasticity (elastic deformation)

the extent of the ability of materials to resume their original shape after stresses of **tension** or **compression** have been applied to them. Materials such as rubber have a high degree of elasticity, and can return to their original shape very easily, but even rubber will eventually deform, beyond its **elastic limit**.

✿ hysteresis

the degree to which a strain depends on the history of previous stresses as well as the present stress. If a stress is only partially removed, most materials will show a greater strain than if the remaining stress had been applied to the unstrained object. When the stress is completely removed the material may—or may not—return to its original shape. Hysteresis also applies in magnetism: when some objects are placed within a magnetic field and then removed, some residual magnetism will remain in the object.

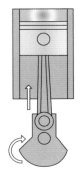

compression stroke

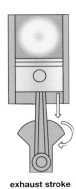

exhaust stroke

The compression ratio in an internal combustion engine: the ratio between the volume in a cylinder at maximum (top) and minimum compression (bottom).

A hysteresis cycle shows the effect of removing the magnetic field from an object then reversing it to reduce magnetization to zero. Then magnetization is reapplied, completing the cycle.

elastic limit

the limit to which an elastic material can be stretched by **stress** without causing a permanent change in its shape.

elastic modulus

the ratio of **stress** to **strain** under particular conditions. This usually remains constant as there are some materials that have several distinct elastic phases until the **elastic limit** is reached. Young's modulus of elasticity refers to tensile testing. It is defined as the applied stretching force per unit area divided by the increase in length per unit length. The shear modulus of **elasticity** (or the rigidity modulus) is the shearing force per unit area divided by the angle of twist. The bulk modulus of elasticity is the compressive force per unit area divided by the change in volume per unit volume.

Poisson's ratio

under conditions of tensile **stress**, the ratio of the transverse contraction to the longitudinal stretching when a tensile stress is applied to a material. A material will get longer as it is stretched and simultaneously become narrower. The ratio varies for different materials. e.g., for lead it is 0.42 and for tungsten it is 0.28.

ultimate tensile strength

the point at which a material or an object under tensile stress actually breaks. There are corresponding terms, "ultimate tensile stress" and "ultimate tensile strain," which are the values of **stress** and **strain** at the point of breaking. These values are normally calculated simply by taking a "before" and "after" reading—what is known as the "engineering stress" and "engineering strain," which differ from the "true" values which take into account the progressive deformation of the material as it is being tested.

creep

the slow deformation of material or an object under constant stress. High temperatures can exacerbate creep in some materials, which should therefore be avoided in constructions likely to be subject to high temperatures. There are normally three stages of creep: "primary creep," during which the strain or deformation increases rapidly with time; "secondary creep," with a slow increase in strain over time; and "tertiary creep," when the rate of creep increases again and breaking point is eventually reached.

Griffith crack length

the limit of the length of a crack in a material that can be tolerated before catastrophic failure occurs. Different materials have different Griffith crack lengths. In a large construction, the longer the crack length the better, because longer cracks are easier to spot and areas of potential danger can therefore be eliminated. There is a theory that the *Titanic* was built of an alloy of steel that had a short Griffith crack length, and thus was structurally weak before it hit the iceberg. Named after British engineer A.A. Griffith (1893–1963), who also stated that if the total energy fell as the result of a crack, the crack would propagate.

stress concentration factor

the ratio of actual stress at a given point in a piece of material to the nominal applied tensile stress of the material. Tensile stress applied to a material increases at points of weakness such as microscopic flaws or cracks in the material, but also where the material has manufactured breaks, e.g., notches, holes, or screw threads. Stress is concentrated at such points.

‍ Moh hardness scale

A rather crude but practical method of comparing the hardness or resistance to scratching of various materials, invented by German mineralogist Friedrich Moh (1773–1839). Despite the name, it is not a scale as such, but lists 10 materials in order of hardness. Each mineral is tested by scratching another material on the list: if it can scratch it, it is harder; if it is itself scratched by the material, then it is softer.

Vickers hardness test

a test for **hardness** using a square-based, pyramid-shaped diamond that is pressed into a metal. The force is applied using a load that can vary between 1 kg and 100 kg, and it is applied for 10 to 15 seconds. The size of the impression made is an indication of the metal's hardness—the smaller the impression, the harder the metal.

‍ Shore hardness test

a test of the hardness of rubber and plastic materials. It uses a hardened indenter which is pressed into the material and the depth of penetration is then measured. The Shore test uses a device called a Durometer and is sometimes also called the Durometer Test. There are a number of different Shore scales, depending on the type of material to be tested.

Charpy impact test

a rudimentary test for impact toughness, still used because it is an economical indicator useful in quality control decisions. It is generally used for assessing the toughness of metals, but similar tests are used for other materials. In the test, a pendulum is used to break by one blow a piece of metal, notched in the middle and supported at each end. The energy absorbed is measured in joules, and this absorbed energy represents the impact strength of the material. The similar Izod test also uses a pendulum but the sample is clamped at the bottom, and the pendulum strikes the top.

The number allocated to each material on the Moh hardness scale does not represent any relationship with the others on the list.

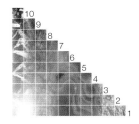

10	diamond	5	apatite
9	corundum	4	fluorspar
8	topaz	3	calcite
7	quartz	2	gypsum
6	feldspar	1	talc

165

The Shore hardness test uses a Durometer. The scale shows the amount of penetration of the material.

❧ bend test

a test usually performed on metal to determine whether it has enough ductility to bend without breaking. It can be used on sheet, strip, plate or wire, and different criteria apply in each case, but generally a standard specimen is bent through a specified arc and, in the case of strip, the direction of grain flow is noted and whether the bend is with or across the grain. The specimen is usually bent over a specified diameter through a specified angle for a specified number of cycles. The common forms of bend test are 3 pt. with two points of pressure on one side, and one between them on the other, and 4 pt. with two points of pressure on each side.

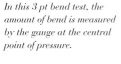

In this 3 pt bend test, the amount of bend is measured by the gauge at the central point of pressure.

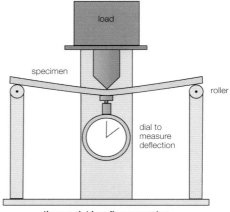

three-point bending apparatus

friction

the force between two materials in contact with each other impairing their ability to slide or roll against each other. The coefficient of friction is a measure of this force, and is very low for two materials that slide together easily, and very high for two that strongly resist movement against each other. The size of the area of contact between the two materials does not matter.

tribometer

a device for measuring the force of resistance between two materials sliding or rolling against each other. In other words, a friction tester.

S.A.E. oil grades

the various grades of lubricating oils. The system was devised by the American Society of Automotive Engineers (S.A.E.), and classifies oils in terms of their viscosity. A light oil is 10 S.A.E. and a heavy oil 40 S.A.E. Most oil sold today is "multigrade" or "multi-viscosity," which is lower viscosity base stock oil blended with a Viscosity Index Improver. V.I.I.s allow a lower viscosity oil. such as a 10 S.A.E. oil. to flow like a 10 S.A.E. oil at low ambient temperatures (for instance. during cold starting) and also flow like a 30 S.A.E. oil at higher ambient and operating temperatures. This example would produce a 10W-30 S.A.E. multigrade oil (the "W" stands for "winter").

pounds per square inch (psi)

the imperial or Customary (**avoirdupois**) measure of pressure or stress. One psi equals 6.895 kilopascals. This measure is often still used in the U.S.

base box

a unit measuring the thickness of tinplate or another form of metallic coating. like galvanizing. It is equivalent to the area formed by 112 sheets of metal. 14×20 inches. or 31.360 square inches. The weight in pounds of one base box is the base weight.

cost-effectiveness

making economical use of resources and techniques: achieving the desired result at the lowest cost. In engineering. the choice of materials is crucial in estimating costs for a project but the cheapest materials are not necessarily the most cost-effective. Materials and equipment must be suitable. otherwise they may fail and further costs will be incurred. The level of energy consumption is also very important and safety factors are crucial. Research and innovation can ultimately cut costs. even if the initial outlay seems high. Often there are several methods of achieving a result. and a cost-effectiveness analysis determines which is chosen.

tire sizes

the markings on the tire wall. giving information about the size of the tire and the wheel. These are now standardized. and the main information is in the form 215/65 × 15. where 215 is the size of the width of the tire in millimeters. 65 is the aspect ratio (the height to width ratio) and 15 is the size of the wheel in inches. The former U.S. and imperial sizes are rarely seen today.

🍂 drill sizes

there are a huge number of drill sizes. in both metric and imperial systems. The imperial system has a series of numbers from 1 to 80. beginning at 1 for a size of 0.2280 inch. and getting progressively smaller. with 80 being 0.0135 inch. Larger sizes have letters. beginning at A for 0.2340 inch. Metric sizes are in millimeters.

The huge array of drill sizes is necessary because of the need for precision in some applications.

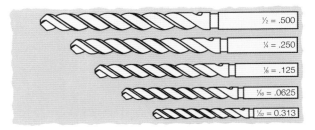

½ = .500
¼ = .250
⅛ = .125
¹⁄₁₆ = .0625
¹⁄₃₂ = 0.313

finance, coins and currencies

✰ modern currencies

the most widely used currencies are the U.S. dollar, the euro, the rouble, the yuan, the yen, the rupee and the pound sterling. Internationally, the U.S. dollar is used as a trading currency of convenience, as it is a very safe currency, backed by one of the world's strongest economies.

✰ pre-decimal British currency

prior to the introduction of decimalization in 1971, the British currency, the pound, was divided into 20 shillings (denoted by the initial "s"), and each shilling was divided into 12 pence (denoted by the initial "d," from the Latin *denarius*). There had also been further subdivisions of the penny, into halfpennies and farthings (a quarter of a penny). Notes were issued for 10 shillings, 1 pound, 5 pounds, 10 pounds and higher denominations. Historically, a gold 1-pound coin, known as a sovereign, had also been issued. Another former coin, the guinea, with a value of 1 pound and 1 shilling, was often still used in prices quoted by shops right up to decimalization.

Roman coins

Roman coins were issued for use throughout the vast Roman Empire, and some remained in use well into the Empire's decline. They were also used outside the Empire as a useful means of exchange. The emperor Augustus standardized coinage, minted in seven denominations (namely the *aureus*, *denarius*, *sestertius*, *dupondius*, *as*, *semis* and *quadrans*), using four different metals—gold, silver, brass and copper.

Greek coins

unlike Roman coinage, ancient Greek coins were not standardized because the various city states issued their own currency, in different designs. The gold *stater* was one basic unit. There were silver *staters*, and often the name *stater* was given to the chief (silver) *drachm* coin issued in a particular state. *Drachm* coins came in various multiples, e.g., the *tetradrachm* was 4 *drachma*. *Drachma* were too large in value to be practical. Similarly, the *obol* was too large in value (and too small in size) to be of much use in everyday transactions.

Principal currencies of the world

Country	Currency
U.S.A.	dollar, cent
Canada	Canadian dollar, cent
Australia	Australian dollar, cent
New Zealand	New Zealand dollar, cent
U.K.	pound sterling, penny
France*	euro, cent(ime)
Germany*	euro, cent
Italy*	euro, cent
Spain*	euro, cent
Russia	rouble, kopek
China	yuan, jiao, fen
Japan	yen
India	rupee, paisa
Indonesia	rupiah
South Korea	won
Israel	new shekel, agora
South Africa	rand, cent
Brazil	real, centavo
Argentina	peso, centavo

Note: several other countries in Europe also use the euro, and many plan to

Pre-decimal British coins and notes and their nicknames

farthing	¼ penny
halfpenny ("ha'penny")	½ penny
one penny	1d
threepence ("thruppence," "threpence", "threpenny bit")	3d
sixpence ("tanner")	6d
one shilling ("bob")	1s, 1/-
florin ("two bob")	2s, 2/-
half-crown ("two and six")	2s 6d, 2/6
crown	5s, 5/-
ten-shilling note ("ten bob note")	10s, 10/-
one pound note ("quid")	£1
five-pound note ("fiver")	£5
ten-pound note ("tenner")	£10

cash

the original low-value Chinese coin, with a square hole in the center, which first appeared in the fourth century B.C.E. and was regularly issued over thousands of years. Larger denominations of cash were also produced: 2, 5 and 10 cash. The Chinese name for the coin is the *tsien*. The word also refers to Indian and Indonesian low-value coins, though there is some dispute about a link with the English word "cash," meaning "ready money."

krugerrand

a South African coin used for investment only, first issued in 1967. It contains one **troy** ounce of gold. It was named for Paul Kruger, first president of the Republic of South Africa, whose head appears on the coins. In 1980, the South African mint began issuing fractional krugerrands, containing $1/2$, $1/4$ and $1/10$ ounce of gold.

louis d'or

often shortened to simply "louis" (and sometimes known internationally as a "pistole"), a former French gold coin worth 10 *livres*. Originally a coin issued in 1640 in the reign of King Louis XIII, it continued to be struck until the Revolution in 1789.

napoleon

a coin originally issued in the time of Napoléon Bonaparte but also issued during the reign of his namesake Napoléon III in the Second Empire. It was a gold coin, worth 20 francs.

piece of eight

a coin of colonial-period America, both North and South, whose name comes from the anglicized form of *peso*, a word derived from the Spanish *pesa* meaning weight, and the fact that there were 8 *reales* in a *peso*. In North America it was also known as a dollar.

thaler

a silver coin issued in Germany, Austria and Switzerland. The word is short for Joachimsthaler, after Joachimsthal, a town now in the Czech Republic, where the coin was first minted. The word "dollar" comes from thaler. Maria Theresa thalers, named for the Austrian empress (1717–80), continued to be minted after her death (still with the date 1780), because they were used in trade with parts of the Ottoman Empire. They are still minted today and are used in some places as a trusted means of exchange.

groat

a British silver coin that was worth four (old) pence, last issued in 1662. The name is of Germanic origin and means "great," on account of the rather bulky appearance of the first coins. It was used to describe several other European coins, and the word *groschen*— the name for the minor value coins in Austria—has the same root.

florin

the name for the 2-shilling piece in pre-decimal British currency. It was a silver, later cupronickel, coin worth 10 new pence, and for a period after decimalization 10 pence coins and florins were both in circulation. It was ultimately replaced in 1993 by a coin half the size. Florin was also an alternative name for the former Dutch currency, the guilder. The original florin was a gold coin from Florence in Italy and also the name of other British and Austrian coins.

talent and mina

the *talent* was an ancient currency unit, and is familiar from Bible stories, but it was also used in ancient Greece. There were 60 *mina* to a *talent*. Both were also units of weight.

sequin

otherwise known as *zecchino*, a name given to various gold coins of Italian city states, Malta and Turkey. They were particularly bright, hence the current use of the word to describe the tiny discs attached to cloth.

❧ ECU

the acronym of the European Currency Unit, before the birth of the euro. The ECU had been a composite currency, whose value was based on a weighted average of different currencies' values. Historically, a number of French coins had been named *écu*, and perhaps partly for this reason it was not retained as a name for the European currency—a pity, because the pronunciation of "ecu" could be fairly standardized across member-states, whereas the name that was adopted, euro, is open to different pronunciations.

National currency weights to the ECU value		
Mar 79– Sept 84	Sept 84– Sept 89	Sept 89– Dec 99
BEF 9.64%	8.57%	8.183%
DEM 32.98%	32.08%	31.955%
DKK 3.06%	2.69%	2.653%
ESP -	-	4.138%
FRF 19.83%	19.06%	20.316%
GBP 13.34%	14.98%	12.452%
GRD -	1.31%	0.437%
IEP 1.15%	1.20%	1.086%
ITL 9.49%	9.98%	7.840%
LUF -	-	0.322%
NLG 10.51%	10.13%	9.98%
PTE -	-	0.695%

scrip

a certificate used during share dealing. It also means the shares released under a bonus issue. The term is sometimes additionally used for paper money that is not regarded as official currency—such notes might be issued during wartime, or in conditions of hyperinflation when previously existing banknotes quickly become obsolete. Small-value notes in the U.S., worth less than a dollar, were also known as "scrips."

gold standard

a system of linking the value of currency to an amount of gold, once common but no longer used. A dollar or pound, for instance, had a specified value in gold. The system had more relevance when coins had a significant gold content. However, most countries have large gold reserves, which can sometimes be used as a means of transferring money to another country. The removal of the dollar from the gold standard and subsequent devaluation led to the collapse of the Bretton Woods system of exchange rates in 1971–72.

exchange rate

the rate at which one currency can be traded for another. In practice, buying and selling rates are slightly different, and banks and bureaux de change generally charge a commission for a transaction. The Bretton Woods agreement controlled exchange rates in the post-Second World War period until 1972, but the widely differing strengths of the world's economies had by then made the system untenable. Exchange rates are now "floating," i.e., they are allowed to find their own level against other currencies, though of course many nations have central banks which step in to either buy or sell currency to slow down depreciation or appreciation.

ஃ consumer price index

a "basket" of commonly bought items, whose changing values when added together can give an idea of changes in the cost of living for the average household in an economy. Closely linked to the inflation rate, although the latter may encompass items that do not affect all households, such as mortgage interest rates.

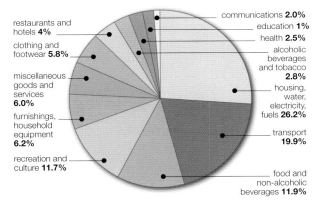

This chart shows a typical breakdown of spending for a household, changes in which indicate increases in the cost of living.

Treasury bill (T-bill)

a short-term bill issued by the U.S. and Canadian governments in various denominations. Treasury bills carry no interest and are tradable on a discount basis with terms to maturity of 3, 6 or 12 months. The difference between purchase price and the face amount represents the return to the investor.

Exchequer bill

an interest-earning bill issued by the British government as a way of raising funds for a contingency, such as war. Exchequer or Treasury bonds for a limited period of time and at fixed interest are still often issued by governments as a way of raising funds for a specific purpose. Gilt-edged securities, or gilts, are another form of government-issued investment, considered very safe and with a fixed rate of interest.

prime rate (minimum lending rate)

the lowest rate of interest charged by major banks in the U.S., being the interest rate offered to their "prime customers." An analogous term in Britain was the minimum lending rate, though this was replaced in 1981 by the less formal "base" rate.

Gross Domestic Product (G.D.P.)

the G.D.P. of a country is the annual total value of all goods and services created domestically within the economy, and so excludes any income from abroad.

⁌ Gross National Product (G.N.P.)

the G.N.P. of a country is the annual total value of all goods and services created by the economy, including investment income from abroad. It is equal to Gross National Income. When calculating G.N.P., it is important to recognize the possible distorting effects of **inflation**. G.N.P. is about output not prices, and a true picture of revenue can only be arrived at by measuring costs of production as well as a rise in prices.

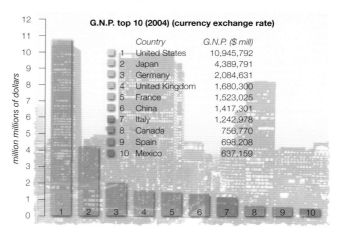

G.N.P. top 10 (2004) (currency exchange rate)

	Country	G.N.P. ($ mill)
1	United States	10,945,792
2	Japan	4,389,791
3	Germany	2,084,631
4	United Kingdom	1,680,300
5	France	1,523,025
6	China	1,417,301
7	Italy	1,242,978
8	Canada	756,770
9	Spain	698,208
10	Mexico	637,159

A comparison of major countries' G.N.P.s. China's economy is growing fast, and it looks set to climb up the table.

trade balance

the difference in value between total exports and total imports over a given period of time. Both exports and imports can be split between "visibles," i.e., tangible goods and products, and "invisibles," i.e., the output of the service sector. If total exports exceed imports, there is a trade surplus: the opposite is a deficit.

money supply

the total amount of money circulating in an economy at a particular time. Within that broad definition, money can be defined in different ways. In Britain, for instance, the two main definitions are M0—notes and coins in circulation outside the Bank of England, cash held in bank (and building society) tills, and banks' operational deposits with the Bank of England; and M4, which is M0 plus all sterling deposits held by parts of the private sector that are not banks or building societies. Control of the money supply is one tool in dealing with **inflation**.

❧ inflation

the general rise in prices of goods and services over a fixed period of time, usually measured over a year, or the fall in the value of money over the same period. If prices are 3 percent higher today than at the same time last year, inflation is said to be 3 percent. Because it is an average, inordinate increases in the prices of specific commodities can distort the picture, and some people's buying power can be affected much more than others. Naturally, in assessing the effects of inflation, increases in earnings over the same period also need to be taken into account. In a situation of fixed exchange rates, inflationary pressure can lead to an official devaluation of a currency.

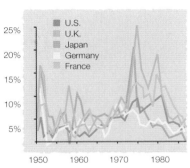

The inflation rates of these five countries varied considerably over the decades. The "oil crisis" of the 1970s affected Japan severely, but it recovered quickly.

depreciation

the fall in the value of a currency against other currencies as the result of an excess in supply of that currency. Depreciation makes imports more expensive and exports cheaper, and in theory an equilibrium will be reached with demand for the currency increasing as the country's exports become more desirable. Unfortunately, inelasticity of demand for exports can mean that this is slow to take place. The term depreciation is also used in accounting when describing the fall in value of a company's equipment because of obsolescence.

duty

a **tax** on certain goods, particularly those imported into a country. In some countries, the relatively high tax on imported alcohol and tobacco ensures the popularity of "duty-free" shops for international travelers. Each country allows a certain amount of duty-free goods to be brought into the country, and the duty-free shops at ports, airports, etc., are able to make a profit by selling travelers their allowance.

tax

a government-imposed charge, designed to raise revenue to fund spending plans. It is usually levied on income from work, sales of goods or services, etc., or based on the value of assets. Income tax is a direct tax, charged at different rates depending on the level of income; sales taxes are paid indirectly when a purchase is made; and property taxes are often charged at a local level to fund the provision of services in a community.

tithe

historically the contribution of part of the produce from farming or a portion of the income of those not involved in farming, made for the upkeep of Church and clergy in a community. The roots of the word are Germanic, meaning "tenth part," because originally it was a tenth of produce or income that was provided.

blood money (wergild)

the value placed on human life in Anglo-Saxon and early medieval Germanic communities. It amounted to compensation for the loss of life paid by the killer (whether the killing was deliberate or accidental) to the victim's family. Both sides then accepted that justice had been done, and no further redress was sought or offered. The amount paid varied according to the status of both killer and victim.

capital

the money that a company is able to draw on to keep it in business, or to start a business. In the latter case, it is called venture capital or risk capital, and is provided by individuals or groups in exchange for shares in the new enterprise. The term capital is also used to describe the net value of a company (after deductions have been made for taxes, overheads and wages).

The difference in calculation between simple and compound interest.

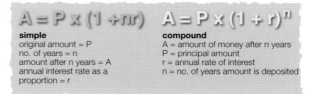

$$A = P \times (1 + nr)$$
simple
original amount = P
no. of years = n
amount after n years = A
annual interest rate as a
proportion = r

$$A = P \times (1 + r)^n$$
compound
A = amount of money after n years
P = principal amount
r = annual rate of interest
n = no. of years amount is deposited

❧ interest

the charge made for borrowing money, often fixed, but sometimes varying in response to changes in the base rate, or over time. Conversely, interest is also the amount paid to investors in return for the use of their money by a government, bank or other institution at a fixed rate or again varying according to changes in the base rate. Compound interest refers to the cyclical reinvestment of money earned in interest, so that the amount on which interest is earned increases each cycle. (However, in the same way, compound interest is also accrued on a debt.) If interest is received only on the initial investment, it is known as simple interest.

Annual Percentage Rate (A.P.R.)

the annualized version of an **interest** rate expressed as a rate for a shorter period of time, e.g., per month. A monthly compound interest rate of 2 percent equals an A.P.R. of 26.24 percent. A simple A.P.R. is calculated by multiplying simple monthly interest by 12.

credit rating

the perceived ability of an individual to repay a loan, depending on the size of the loan. It varies according to the size of the person's income or assets, and account is also taken of the person's previous record in repaying loans. Individuals therefore sometimes find it difficult (or even impossible) to get a loan if they haven't had one before, or if they have reneged on payments on a previous loan. Similar criteria may be applied to companies.

earnings

money earned by an individual as reward for work or as interest from investments. It may be stated as the gross amount (before tax deductions) or net (after tax). Earnings also means the profits of an enterprise or different sectors of the economy.

float

a small amount of cash held at the start of trading in a shop till to provide change for initial customers, and hence to be deducted before takings are added up at the end of the day. Also used in the U.S. and Canada to mean minor cash amounts generally (often referred to as "petty cash"), or money due from uncollected checks.

share price

the price of a share in a company's equity. It is the price that a potential investor will be quoted, and as with prices of goods, is partly based on true market value, while taking into account the need to attract buyers. So, sometimes shares will be sold at a lower price than their real worth in order to bring new investment into a company. A rights issue is an offer made to existing shareholders to buy further shares at a discounted price.

American Depository Receipt (A.D.R.)

a block of shares held in foreign corporations, issued in the form of a certificate by a U.S. bank, that can be quoted, bought and sold on the stock market as any other shares. Investors wishing to invest in foreign corporations contact a bank (via their broker) and A.D.R.s are issued once the transaction is finalized.

yield

the return on an investment in shares or a gilt or bond. For shares, yield is measured by taking the annual **dividend** and dividing by the **share price**. This calculation can cause problems for investors in assessing a company's prospects: a low yield may indicate that the share price is high because the market views the company's prospects favorably, or that the dividend is low because the company is performing badly.

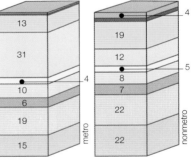

agriculture, forestry and fishing
mining
manufacturing
producer services
recreation
transportation, utilities and wholesale
construction
consumer services
government

175

Percentage earnings in different sectors of the U.S. economy in 2001, in metropolitan and nonmetropolitan areas.

dividend

a way of distributing part of a company's profits among share-holders, usually paid annually or half-yearly. The dividend is quoted on a per share basis, in order to make transparent the amount of profits being paid to shareholders and the amount being retained for reinvestment. New companies will not be expected to pay a dividend in the first few years of their existence, because they need to plow back all profits into the business.

price–earnings ratio

the ratio between the value of a share and the earnings from it, calculated by simply taking the market value of the share and dividing by the earnings per share. It is considered the best way of evaluating the performance of shares. A "trailing" price–earnings ratio looks at performance over the past four quarters, and gives a truer picture of overall performance than if one quarter is used. A "forward" ratio is an estimate of future performance, either by the company itself or an independent analyst. Potential investors have to examine both ratios in order to determine the advisability of investment.

turnover

the total income of an enterprise, usually measured annually, and without deducting outgoings on overheads or investment in new plant, etc. The term is also used to describe the rate at which stock in trade is sold and replaced, and the rate at which members of the workforce leave and are replaced.

income

all money received by an individual in the form of wages or from investments. It is all subject to income **tax**. The term is also used to describe total revenue received by a firm (and as such is a synonym for **turnover**), or the revenue of even larger institutions or economic units, including countries, whose income is related to the **Gross National Product**.

profit

the difference between the amount of **turnover** of an enterprise and the costs of producing its products or providing a service. Gross profit takes no account of overheads, depreciation, the amount paid in wages or **interest** payments, whereas net profit, after these deductions, is the figure usually quoted as a firm's profit (before or after **tax**). Operating profit is a firm's profit from normal trading activities, and is calculated by subtracting direct and indirect costs from the trading profit. In pure economics, normal profit is the profit that a company needs to make from normal trading activities to keep it from using its means of production in other activities.

loss

negative profit, when the amount of **turnover** of an enterprise is outweighed by the costs of producing its products or providing its services. In pure economic theory, making a loss causes a firm to exit the industry, but in practice there are circumstances when a loss is regarded as acceptable, particularly when new firms enter an industry, and need time to establish a customer base.

margin

the **profit** margin of an enterprise is its operating profit expressed as a percentage of its turnover. For example, a company with a turnover of $4 million and an operating profit of $500,000 would have a profit margin of 12.5 percent. The operating margin is the turnover of the enterprise minus direct costs and overheads.

❧ unit cost

the cost of producing a single item for sale, measured by taking the total cost of the production run and dividing by the number of items.

overheads

the costs of an operation that are not directly linked to the actual production of goods or services, sometimes referred to as indirect costs. These are the fixed costs of rent, maintenance contracts, wages and other expenses. Overheads are not necessarily fixed, however, as, for example, extra machinery may need to be rented to increase production, or extra wages may be paid in the form of overtime.

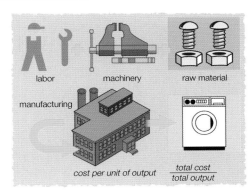

labor   machinery   raw material

manufacturing

cost per unit of output $\dfrac{\text{total cost}}{\text{total output}}$

177

The various inputs that are brought together in the manufacturing process determine the unit cost.

working capital

the assets of a company available for use by the business in improving its turnover. Buildings that are used for production cannot therefore be considered working capital, as they are not available as a means of raising revenue. Another definition of working capital is assets that are not offset by liabilities. In other words, assets that are a source of revenue can only be used as capital if liabilities are dealt with first. It is possible to have negative working capital though this situation clearly cannot be sustained for long, as companies with more assets than liabilities always tend to outperform companies whose assets are outweighed by liabilities. Another name for working capital in this sense is liquid assets.

value added

the value of an item, product, or even entire enterprise at the end of a set period minus its value at the beginning of the period.

The term can also be used when describing the activity of an entire country's economy, in which case the value added amounts to the **Gross Domestic Product**. A value-added **tax** is an indirect tax that is constantly passed on through each stage of the production process. Hence, a supplier is able to deduct tax paid for materials from the tax received from a buyer.

goodwill

a somewhat indefinable term recognizing economic benefits accruing from networking, loyalty, commercial reputation, and "goodwill gestures." These benefits are almost impossible to quantify, yet sometimes withdrawal of goodwill can have serious consequences. A company may lose a contract not because it suddenly becomes uncompetitive, but because any edge that goodwill gave it over its rivals has now gone.

book value

the value of a company's assets as shown in its published accounts or "books"—or the real value of a company calculated from the value of its book assets minus its book liabilities. Generally, book value is the very basic value of a company, and its "real" value as far as investors are concerned is often much higher.

Net Asset Value (N.A.V.)

the total assets of an enterprise after its liabilities and capital charges have been paid for. In this definition, it is therefore essentially the same as **book value**. In the U.S., the N.A.V. is the value of one mutual fund share, which is calculated by dividing the net assets of the fund by the number of shares outstanding.

liquidity

the amount of assets owned by an enterprise that can easily be converted into money (usually for **working capital**). The liquidity ratio is the ratio between liquid assets and the total assets of a bank or other financial institution. The related term "liquidation" refers to the termination of a business by realizing the value of assets in order to cover liabilities.

The number of bankruptcies in a particular area will inevitably be influenced by local market conditions.

number of bankruptcy cases

159,700
25,000–51,300
15,000–25,000
5,000–15,000
900–5000

Source: Administrative Office of the U.S. Courts, 1993

❧ bankruptcy

the judgment by a court that a person or company is insolvent. Any property is then transferred to a trustee, who as far as possible will settle debts using the assets of the person or company. The trustee thus acts a liquidator (see **liquidity**).

food

spoon sizes

measuring spoons come in various sizes, but the most common are
$1/4$ teaspoon, $1/2$ teaspoon, teaspoon, dessertspoon and tablespoon.
Generally in liquid measures, 1 teaspoon = 5 ml ($1/6$ U.S. fl oz), 1
dessertspoon = 10 ml ($1/3$ fl oz) and 1 tablespoon = 15 ml ($1/2$ fl
oz), though there is no universal agreement on these definitions.
With dry measures, 1 tablespoon = $1/2$ oz (14.235 grams).

spoon measures

a level spoonful has the excess scraped off with a straight knife so
that the ingredient is literally level with the edge of the spoon. A
rounded spoonful has about as much above the level of the edge
of the spoon as within it. A heaping or heaped spoonful is as much
as the spoon will hold.

stick

a stick of butter is equivalent to 8 tablespoons ($1/2$ cup, 4 oz, or
125 grams). In the U.S., butter usually comes in a pound box, divided into four "sticks." The wrapping also shows tablespoon and
teaspoon measures.

cup

for liquid measures, 1 cup = 8 U.S. fl oz (237 ml, rounded up in
recipes to 250 ml) or $1/2$ U.S. pint. In North America, a cup is the
equivalent of 250 ml for dry ingredients.

handful

a vague measure which only really applies to dry ingredients, but
a rounded handful is sometimes described as a quantity looking
like half a baseball. This amounts to $1/2$ cup. A level handful is
$1/3$ cup. A handful is only really used for ingredients whose quantities do not need to be exact, and where it is a matter of taste how
much to add.

knob

just what is a "knob of butter"? The expression is gradually losing
currency as people demand more precise measurements in recipes.
A "knob" is another of those terms that allows you to decide how
much you think ought to go in. One or two tablespoons at most.

dash

for liquids, a dash is defined as 6 drops, 76 drops making 1 teaspoon or $1/6$ U.S. fl oz (5 ml), so it's a very small amount. But it can
also refer to a small quantity of something dry that has a powerful
flavor, like spices. The use of qualified expressions, like "a gener-
ous dash," show that it is all a matter of taste.

tablespoon dessert- teaspoon
15ml   spoon   5 ml
            10ml

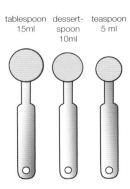

*There are various spoon
sizes available. These are
the relative sizes of a
tablespoon, a dessertspoon
and a teaspoon.*

pinch

a pinch is a small amount of a dry ingredient. It is well known in the phrase "take it with a pinch (or grain) of salt," meaning "treat it skeptically." Why adding salt should be the solution when faced with something of doubtful authenticity is hard to fathom, but it seems to suggest that having a little salt will improve the flavor—making it seem better than it really is. Too much salt obviously will spoil it. Common sense suggests that a pinch is literally just that: no more than you can get between thumb and forefinger, sometimes defined as $1/8$ teaspoon or less.

ε milk and cream types

there are three main types of milk: whole milk; skimmed, skim or non-fat milk; and semi-skimmed or low-fat milk. Homogenized milk has been treated so that the cream remains distributed throughout the milk; otherwise, it will separate and come to the top. Pasteurized milk has been heat-treated to kill harmful micro-organisms. U.H.T. (ultra high temperature) pasteurization sterilizes the milk, allowing it to keep much longer, but significantly alters the taste. The main types of cream are heavy cream (double in the U.K.), whipping cream and light or single cream. One other variety of cream worth mentioning is sour or soured cream, which has had a natural culture added to it.

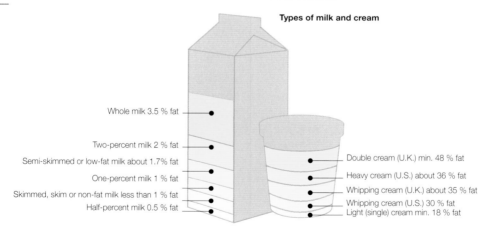

Types of milk and cream

Whole milk 3.5 % fat

Two-percent milk 2 % fat

Semi-skimmed or low-fat milk about 1.7% fat

One-percent milk 1 % fat

Skimmed, skim or non-fat milk less than 1 % fat

Half-percent milk 0.5 % fat

Double cream (U.K.) min. 48 % fat

Heavy cream (U.S.) about 36 % fat

Whipping cream (U.K.) about 35 % fat

Whipping cream (U.S.) 30 % fat

Light (single) cream min. 18 % fat

egg sizes

in the U.S. size is determined by minimum weight per dozen and in Canada by weight per egg. The North American sizes are pee wee (15 oz/less than 42 grams); small (18 oz/42 grams); medium (21 oz/49 grams); large (24 oz/56 grams); extra large (27 oz/68 grams); and jumbo (30 oz/70 grams). In Britain, egg sizes are small (up to 52.9 grams); medium (53–62.9 grams); large (63–72.9 grams); and very large (73 grams and over). Eggs are also graded. In the U.S., AA are the best, followed by A, B and C. In Britain, there are two grades: A, which are sold as eggs, and B, which are broken out and pasteurized.

grades of sugar

granulated sugar is the standard form, with small white sugar crystals. Caster, fine or extra fine sugar has even smaller granules, and confectioner's or icing sugar is ground to a fine powder, used to make icing and for dusting cakes. Sugar cubes are granules stuck together with syrup in small blocks. Brown sugars come in various forms, mostly with smaller crystals than white sugar, and contain molasses. Muscovado or Barbados sugar has a strong molasses flavor and Demerara has golden, slightly sticky granules. Liquid sugars include golden syrup, treacle, corn syrup and maple syrup.

⁊ sugar cooking stages

the different stages in the cooking of sugar with water to make syrup. The higher the temperature, the more water evaporates, and the harder the syrup will set when it cools. Each stage has a name and each is separated by only a few degrees.

unit (of alcohol)

a term to describe a measure of alcohol, used particularly in health warnings. A unit is 8 grams or 10 ml of pure alcohol. The number of units in a drink is defined as its volume in milliliters multiplied by the percentage of alcohol by volume (% A.B.V.) and divided by 1,000. Thus 500 ml of beer with a 5 percent A.B.V. is 2.5 units. Safe levels of alcohol consumption are defined as a certain number of units per day, fewer for women than for men.

% A.B.V. (alcohol by volume)

the proportion of alcohol in a drink, expressed in percentage terms. One fairly reliable way of calculating it is to take a reading of the original specific gravity before fermentation using a hydrometer, measuring again once fermentation is complete. The A.B.V. is the difference between the two readings divided by 7.36. In the U.S., the A.B.W. (alcohol by weight) figure is sometimes used. To convert A.B.W. to A.B.V., multiply by 1.267.

% proof (degrees proof)

proof refers to the amount of alcohol in a drink. To describe the proof number as a percentage is misleading, as the percentage of alcohol is actually half the proof number. So, in the U.S., 100 proof, or degrees proof, means the drink has 50 percent alcohol. This was originally calculated by adding gunpowder to a drink and trying to ignite it. If it didn't ignite, there was too much water in it. It was thought that the gunpowder would ignite as soon as the percentage of alcohol reached 50 percent. The correct figure has

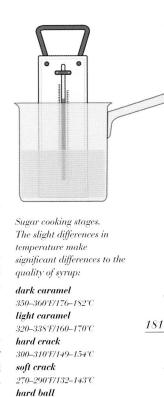

Sugar cooking stages. The slight differences in temperature make significant differences to the quality of syrup:

dark caramel
350–360°F/176–182°C
light caramel
320–338°F/160–170°C
hard crack
300–310°F/149–154°C
soft crack
270–290°F/132–143°C
hard ball
250–265°F/121–129°C
firm ball
242–248°F/116–120°C
soft ball
234–240°F/112–116°C
blow or soufflé
230–235°F/110–112°C
thread
223–235°F/106–112°C
pearl
220–222°F/104–106°C

since been found to be 57.15 percent and the British proof system is still based on this. Hence British 100 proof spirit is a lot stronger than its U.S. equivalent.

demijohn

a large bottle with a short narrow neck, often with handles, sometimes encased in wickerwork. These bottles were originally used to transport or store liquids but are frequently used today by home winemakers to ferment the grape juice in. A common size for winemaking is about a gallon or 4.5 liters, but demijohns can be much larger than this; some can contain as much as 20 gallons—or more.

thimbleful

a thimbleful, as the word suggests, is a very small quantity of a drink (or other substance). Often used as a response when offered a drink: "Just a thimbleful"—but probably more is expected than the phrase suggests!

shot

a shot is defined (vaguely) as "a small amount of drink," and usually applies to spirits or hard liquor. It is sometimes defined in the U.S. more precisely as the amount of liquor that can be held in a small glass called a jigger, i.e., $1^1/2$ fl oz (44.4 ml).

optic

a device used in bars for accurately measuring drinks which is fixed below an inverted bottle so that the drink flows into it, filling it to the exact measure. It is called an optic because it is made of glass and you can see the liquid as it is dispensed. The liquid is released by holding the glass below the outlet and pushing up against levers. Bar optics have less application in the U.S. because there are no standard measures of spirits (*see* **measure**).

measure (of spirits)

in Britain spirits were once measured in gills (1 gill = $1/4$ U.K. pint). The size of a measure varied regionally, from $1/6$ gill to $1/4$ gill. Now the European Union has imposed standard measures of 25 ml and 35 ml, the latter becoming more common as the measure used in British pubs. Similar rules exist in other countries: in Australia, spirit measures are 15 ml, 30 ml and 60 ml, but no standard measures apply in the U.S.

split

in the U.S., a measure of carbonated water or wine, usually 6 fl oz (177 ml), sometimes a little more. A quarter-bottle of wine, 187 ml, is also often referred to as a split. The term originated from the practice in bars of splitting a bottle between two customers, a 6 fl oz bottle being regarded as enough for a drink each. It is also widely used (apparently erroneously. though that of course is a matter of dispute) to denote a half-bottle of wine (375 ml). because this is a "split" of a whole bottle.

finger

an imprecise measure of whiskey (or other spirits). defined roughly as a depth of $3/4$–1 inch (2 cm) in a glass. about the width of a man's finger. Clearly. the volume depends on how wide the glass is.

glass (of wine)

wine glasses vary greatly in size. but in pubs and restaurants in the U.K. a "small" glass is 125 ml. a "medium" one 175 ml and a "large" one 250 ml. Away from the strictly controlled commercial environment. the size (and shape) of glasses will vary considerably. depending on the wine to be served in them. Red wine glasses are larger than white.

drinking stein

an earthenware beer mug. often with an elaborate design. popular in Germany. A "stein" is also used to describe a quantity of beer. The most common sizes are $1/2$ liter and 1 liter. though ornamental steins come in anything from a miniature version to those containing far too many liters to be seriously thought of as a drinking vessel.

Calorie

a large Calorie is the amount of a specific food able to produce 1.000 (small) calories of energy. This energy is released when the food is oxidized. The large Calorie is a unit frequently seen in diet sheets and on food nutrition labeling: the greater number of Calories a food contains. the more weight the consumer is likely to put on. However. weight gain is also dependent on the expenditure of energy. The small calorie as a unit of energy has largely been replaced in the scientific world by the SI unit—the joule.

❧ R.D.A.

the Recommended Daily Allowance. also known as the Recommended Dietary Allowance. The R.D.A. is defined as the amount of a substance. a vitamin. mineral or protein. that is

The Recommended Daily Allowances of vitamins and minerals. All these substances can be taken as supplements but they also naturally occur in foods.

nutrient	amount
Vitamin A	5,000 international units (IU)
Vitamin C	60 milligrams (mg)
thiamin	1.5mg
riboflavin	1.7mg
niacin	20mg
calcium	1.0 gram (g)
iron	18mg
Vitamin D	400 IU
Vitamin E	30 IU
Vitamin B6	2.0mg
folic acid	0.4mg
Vitamin B12	6 micrograms (mcg)
phosphorus	1.0g
iodine	150mcg
magnesium	400mcg
zinc	15mcg
copper	2mg
biotin	0.3mg
panothenic acid	10mg

suggested for daily consumption in order to maintain good health. These prescribed amounts for each substance vary according to different groups in the population, depending on age, gender, whether a woman is pregnant, etc. The recommendations usually also indicate which foods are the best sources of each substance.

R.D.I.

the Reference Daily Intake. A term recently introduced in the U.S. which replaces **R.D.A.** in voluntary nutrition labeling to describe the amounts of vitamins, minerals and protein contained in foods. In practice, the R.D.I. values are the same as the old R.D.A. values, but the notion of "recommendation" has been removed, as this gave the impression that a certain amount of these substances "should be" consumed every day.

Daily Reference Value

the Daily Reference Values are recommended daily maximum amounts of total fat, total carbohydrate (including fiber), protein, cholesterol, potassium and sodium, established by the U.S. Food and Drug Administration. The figures vary depending on the amount of **Calories** consumed each day.

retinol equivalent (R.E.)

a unit used to measure the activity of different sources of Vitamin A. One R.E. is normally defined as being equal to 1 microgram of retinol, 6 micrograms of beta-carotene or 12 micrograms of other provitaminic carotenoids. Thus, the body needs six times as much beta-carotene as retinol in order to absorb Vitamin A at the same rate. The rate of absorption also depends on other factors such as diet and medical condition. There is, however, some disagreement over the values for the carotenoids.

F.C.C. unit

the Food Chemical Codex unit, used to measure the purity and effectiveness of chemicals added to foods. The F.C.C. is a set of standards prepared by the U.S. Institute of Medicine for the U.S. Food and Drug Administration. It covers such things as the amount of the enzyme lactase, well known to people who are lactose-intolerant. The F.C.C. unit is not an exact amount in milligrams, because of the different ways in which the chemicals are prepared by manufacturers. Rather, it is a measure of effectiveness, regardless of the weight of the substance.

≈ gas mark

the gas mark temperature scale commonly used in stoves in Britain and some Commonwealth countries has precise Fahrenheit and Celsius equivalents. Gas mark 1 will make the stove temperature 275°F (140°C), while gas mark 8 will make it 450°F (230°C).

gas mark	temp (°F)	temp (°C)	description
¼	225	110	very cool
½	250	130	
1	275	140	cool
2	300	150	
3	325	170	very moderate
4	350	180	moderate
5	375	190	
6	400	200	moderately hot
7	425	220	hot
8	450	230	
9	475	240	very hot

Most British recipe books will give temperatures in gas marks as well as in degrees Fahrenheit and Celsius.

liquids

gallon

a unit of liquid measurement in the imperial system, still used in the U.K., Canada and especially the U.S., but slowly being superseded in by the liter. In British and Canadian usage, the imperial gallon represents the volume occupied by 10 lb **avoirdupois** of water, or 4.54609 liters; the U.S. gallon, however, is a smaller volume, equivalent to 3.7854 liters. Both in Britain and the U.S. the gallon can be subdivided into 8 pints, or 4 quarts, or 32 gills, but these measures necessarily have differing respective values in each system. To confuse matters further, the U.K. gallon divides into 160 U.K. fluid ounces, whereas the U.S. gallon divides into 128 U.S. fluid ounces; and though the U.K. gallon is the same in both liquid and dry measurement, the U.S. dry gallon is a separate unit, equivalent to 4.40476 liters.

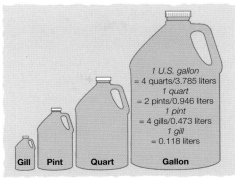

1 U.S. gallon
= 4 quarts/3.785 liters
1 quart
= 2 pints/0.946 liters
1 pint
= 4 gills/0.473 liters
1 gill
= 0.118 liters

Gill   Pint   Quart   Gallon

Probably because they are so commonly in use for everyday purchases such as milk, beer, gasoline, etc., traditional liquid measures are only very reluctantly being replaced in the U.S. and U.K.

quart

a unit of liquid measurement equivalent to a quarter of a gallon, or two pints. In the U.K. system a quart is the equivalent of 1.1365 liters; a U.S. quart equals 0.9463 liters. Now virtually obsolete in Britain and Canada, even among adherents of the imperial system, the quart is still widely used in the U.S.

pint

a unit of liquid measurement equivalent to $1/8$ gallon, $1/2$ quart, or 4 gills. One U.K. pint is equivalent to 0.5683 liters; 1 U.S. pint equals 0.4732 liters. The pint is widely used in the U.S., where there is sometimes fierce resistance to the adoption of metric units, but it also seems likely to remain in common use in the U.K. even after complete metrication has been achieved—no doubt because it is the preferred unit for dispensing milk and, perhaps more importantly, beer. Pints (and fluid ounces) are little used in Canada.

fluid ounce (fl oz)

a unit of liquid measurement, a small subdivision of the pint, but representing different quantities in the U.K. and U.S. Not only is the U.K. pint a different actual volume from the U.S. pint, it is also divided into a different number of fluid ounces: whereas 1 U.K. pint is 20 U.K. fl oz (therefore 1 U.K. fl oz = 28.4131 ml), the U.S. pint is 16 U.S. fl oz (1 U.S. fl oz = 29.575 ml).

❧ gill

a unit of liquid measurement, a quarter of a pint. Like most U.K. and U.S. liquid measurements, the actual volume is different in the two countries: 1 U.K. gill = 0.142075 liters; 1 U.S. gill = 0.1183 liters. Furthermore, the U.K. gill is divided into 5 (U.K.) fl oz, but the U.S. gill is divided into 4 (U.S.) fl oz. As with the U.K. pint, the gill will only reluctantly fall from use in Britain, as spirits are served there in fractions of a gill.

fluid scruple

a small unit of liquid measurement, $1/3$ of a fluid dram, or $1/24$ of a U.K. fluid ounce. This is the equivalent of 1.1839 milliliters. The fluid scruple is even more rarely used than the apothecaries' unit of weight, the **scruple**, from which its name is derived, and does not appear in the U.S. system of liquid measurement.

fluid dram (drachm)

a small unit of liquid measurement equivalent to $1/8$ of a fluid ounce. In the U.K., this means that 1 fluid dram is $1/160$ of a U.K. pint, but in the U.S. it is $1/128$ of a U.S. pint—two very different quantities: in metric terms, 1 U.K. fluid dram = 3.5519 ml, 1 U.S. fluid dram = 3.6969 ml. The term fluid dram is derived from the dram or drachm of the apothecaries' weight system.

minim

a very small unit of liquid measurement equivalent to $1/60$ of a fluid dram. Because of the different quantities and subdivisions of the U.K. and U.S. fluid measures, 1 U.K. minim = 0.0592 ml, and 1 U.S. minim = 0.0616 ml.

cubic meter (m^3)

a unit of volumetric measurement, normally associated with dry goods but sometimes used as a liquid measure. Unsurprisingly, it is derived from the SI meter: it is equal to the volume of a cube with sides of 1 meter, and is the equivalent of 1,000 liters or 1 kiloliter. In more common usage as a liquid measurement is the derived unit cubic centimeter (cm^3 or cc), one-millionth of a cubic meter, which is equivalent to 1 milliliter; engine capacity of motor vehicles is frequently measured in either cc or liters.

liter (litre; L; l)

a unit of volumetric measurement in the SI system, the volume of 1 kg of water at 4°C, for general purposes equivalent to 1 cubic decimeter. It is the SI unit for liquid measurement, but is also often used for dry goods. The abbreviation L was adopted officially in 1979, but l is still widely used. One liter is equal to 1.760 U.K. pints, or 2.1134 U.S. pints. Like all the SI base units, derived units can be expressed by suffixes such as deci-, milli-, deca-, kilo-, etc.

milliliter (millilitre; mL; ml)

a unit of liquid measurement equal to one-thousandth of a liter. For general purposes, it is the equivalent of 1 cubic centimeter (cm^3 or cc), and 1 ml = 0.0352 U.K. fl oz or 0.03381 U.S. fl oz. The abbreviation mL is the officially preferred form, but ml is still widely used; in speech the term "mil" is often heard, but only in contexts where its meaning is clear, as "mil" can also be used to refer to a number of other measurements with the milli- prefix, and even as a short form of "million."

hu

a Chinese unit of liquid measurement, equal to 51.773 liters. It is subdivided into 50 *sheng*.

sheng

a Chinese unit of liquid measurement, $^1/_{50}$ of a *hu*, equivalent to 1.035 liters. In recent usage this has been rounded down so that in the new Chinese system, 1 *sheng* = 1 liter.

❧ seah (se'a)

an ancient Hebrew unit of volumetric measurement, used for both liquid and dry goods, equivalent to approximately 13.44 liters. It was subdivided into two *hin*, and was equal to one-third of a *bath*.

❧ hin

an ancient Hebrew unit of volumetric measurement, equal to half a *seah*, or about 6.72 liters. The *hin* was subdivided into 3 *cab* (or *kab*), or 2 *logs*.

❧ bath (bat; bathos)

an ancient Hebrew unit of volumetric measurement, equal to 3 *seah*, or about 40.23 liters.

❧ kor (cor)

an ancient Hebrew unit of liquid measurement, equivalent to 10 *bath*, 30 *seah*, or about 402.3 liters. The equivalent measure for dry goods was known as a *homer* or *chomer*.

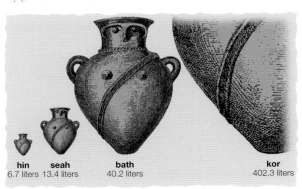

hin **seah** **bath** **kor**
6.7 liters 13.4 liters 40.2 liters 402.3 liters

The ancient Hebrew units of liquid measure were used mainly in the oil and wine trades, but also as a measure of dry goods.

ligula

a small unit of liquid measurement in ancient Rome. $1/24$ of a *hemina*, or about 11.0677 ml.

hemina

an ancient Roman unit of liquid measurement, used particularly in the wine and oil trades. It was equal to half a sextarius, and could be divided into 24 *ligulae*. It was the rough equivalent of the modern half pint, and with similar usage: more exactly 1 *hemina* = 0.2656 liter (0.467 U.K. pint or 0.561 U.S. pint).

sextarius

a unit of liquid measurement common in ancient Rome, roughly equivalent to the modern pint (1 *sextarius* = 0.935 U.K. pint or 1.1227 U.S. pint), and in similar usage for wine and oil. It was equal to $1/6$ of a *congius* (hence its name "*sextarius*," Latin for "sixth"), or 2 *heminae*, approximately 0.531 liters.

congius

an ancient Roman unit of liquid measurement, equal to 6 *sextarii*, $1/4$ of an *urna*, or about 3.1875 liters. The term *congius* was also adopted in Britain for a time in the 19th century as an alternative name for the U.K. imperial gallon in medicine and pharmacology.

urna

an ancient Roman unit of liquid measurement, equal to 4 *congii*, or approximately 12.75 liters.

acre inch (ac in)

a volumetric unit in the measurement of bodies of water such as reservoirs. One acre inch is the volume of water covering 1 acre to a depth of 1 inch, and contains 3,630 cubic feet (approximately 102.79 cubic meters). A related, and widely used measure is the acre foot (af), equal to 12 acre inches and thus equivalent to 43,560 cubic feet (approximately 1,233,482 cubic meters).

million acre feet (maf; Maf)

a measurement of the volume of bodies of water, one million times the size of the acre foot, i.e., 3,630 million cubic feet or approximately 102.79 million cubic meters.

flask

an iron container for the transportation of mercury, holding $76 1/2$ lb (34.7 kg) of the liquid metal. The term is also used loosely for many different-sized bottles and bottle-shaped containers, usually to contain alcoholic drinks, and especially spirits.

droplet

a very small quantity of liquid, literally "a small drop." A drop was originally the unit used in dispensing and administering medicines.

measured by use of a glass dropper, and roughly equivalent to a **minim**, and later taken to mean a minim. A droplet thus came to mean any quantity of liquid, smaller than a drop or minim, capable of forming a spherical shape. In practice, the term is now used for quantities smaller than about 0.05 ml, and normally measured by the diameter of the droplet rather than its volume. The size of droplets is particularly important in fields where liquids are atomized or nebulized, such as medical nasal sprays; or where the consistency of emulsions is critical, such as the food industry.

Clark scale

a scale for measuring the hardness of water, named after the 19th-century scientist Hosiah Clark. The scale is divided into units known as degrees Clark; 1 degree Clark is now taken as 1 part of calcium carbonate per 70,000 of water (derived from the original unit of 1 grain per U.K. gallon), or about 14.3 parts per million.

specific gravity

the ratio of the mass of a given volume of a liquid to the mass of the same volume of water at 4°C (i.e., at maximum density). The more usual scientific term is now "relative density," but specific gravity is used (along with "original gravity," or OG, the specific gravity of the unfermented wort) in the brewing trade to measure the potential alcoholic strength of beer, although this is being superseded by reference to the percentage of alcohol in the finished product.

✥ hydrometer

an instrument for measuring the density of liquids. It usually takes the form of a glass bulb, containing a weight, attached to a calibrated glass stem. The weight floats vertically in the liquid to be measured, and a reading can be taken at the point where the surface of the liquid reaches on the scale. A similar device, known as an oleometer, is used to determine the purity of oils.

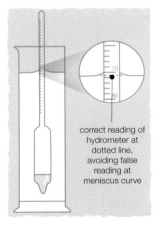

correct reading of hydrometer at dotted line, avoiding false reading at meniscus curve

Most hydrometers are calibrated to indicate the relative density of the liquid being tested, but some specialized instruments are graduated to show the concentration of a particular solution.

hypsometer

an instrument for measuring the boiling point of liquids, usually water. As boiling point is dependent on atmospheric pressure, by comparing a hypsometer reading with boiling point at sea level, the instrument can be used to calculate altitude.

surface tension

the property of liquids that gives them the appearance of having a "skin" in a state of tension, caused by unbalanced attractive forces of molecules in the surface. Surface tension is measured in terms of the force acting on the surface, and at right angles to it, in units of force per unit length (normally newtons per meter—Nm^{-1}). The property is responsible for such phenomena as the ability of small insects to "skate" on the surface of water, and the spherical shape of soap bubbles.

solubility

loosely, the capability of a substance to dissolve in a liquid; more specifically, the amount of a substance that can be dissolved in a liquid at a certain temperature. It can be expressed in terms of mass per unit volume, percentage, parts per million, or moles per kilogram or liter.

miscibility

the capability of two or more liquids dissolving together at a particular temperature to form a mixture.

eutectic

a mixture of substances whose composition is such that the lowest possible temperature for solidification is achieved. In a eutectic (more properly a eutectic mixture) any change in the relative quantities of its constituents would raise the temperature at which it solidifies. If a graph showing temperature in relation to the constituents of a mixture, known as a **phase diagram**, is drawn, the eutectic point is the temperature on the phase diagram that corresponds to the formation of a eutectic. At this temperature, crystallization of the constituent parts occurs simultaneously.

ullage

the amount by which a container falls short of its full capacity. The term is used today mainly in shipping, and represents the spare capacity of a partially filled container (such as a tank), but has its origins in the time when liquids were transported in barrels, and a fair amount of spillage, leakage and evaporation was to be expected. (Whisky distillers in Scotland affectionately refer to this loss as the "Angels' Share.") The amount of this shortfall, known as the ullage, had to be compensated for by the seller. In the transportation of beer, wine and spirits, this was done by refilling the cask with the requisite liquor up to the tap-hole. It is thought that the word derives from the French for this tap-hole, *oeuil* (meaning "eye"), and hence *oeuillage*; another etymology is from the old French verb *oeiller*, meaning to fill.

sea

a large geographical mass of salt water, smaller than an ocean. The majority of the seas are in fact subdivisions of the major oceans, or immediately adjacent to them, but many, such as the Mediterranean and Baltic, are separate and virtually large salt-water lakes. Despite the old sailors' boast of having sailed "The Seven Seas," there are more than 20 bodies of water internationally recognized as seas.

ocean

a huge geographical mass of salt water. The salt water that covers more than 70 percent of the Earth's surface is normally divided into three major oceans, the Pacific, Atlantic and Indian; the Arctic and, more rarely, Antarctic Oceans are sometimes considered to be separate oceans, but geographers increasingly determine these two to be merely the northern or southern extremes of the three major oceans. Estimates of the total surface areas of the oceans vary considerably, in part because of the difficulty of accurate measurement, but often because it is not made clear whether the dimensions include the adjacent seas, gulfs and channels.

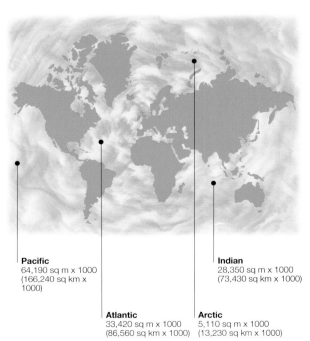

The surface area of the major oceans, excluding the adjacent seas and treating the Arctic Ocean as a separate entity, are indicated in square miles and square kilometers. The average depth of the world's oceans is around 13,100 feet (4,000 m). The greatest depth, 36,160 feet (11,022 m), is in the Pacific Ocean.

Pacific
64,190 sq m x 1000
(166,240 sq km x
1000)

Indian
28,350 sq m x 1000
(73,430 sq km x 1000)

Atlantic
33,420 sq m x 1000
(86,560 sq km x 1000)

Arctic
5,110 sq m x 1000
(13,230 sq km x 1000)

paper and publishing

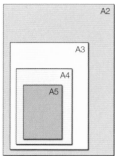

The A series of paper sizes has a constant height:width ratio, as shown above.

ᔆ I.S.O./A.B.C. series

a system of paper sizes adopted by the I.S.O. (International Standards Organization) in Switzerland and now used by most countries outside North America. In the I.S.O./A.B.C. series, the height-to-width ratio of all pages equals the square root of two (1.4142 : 1). Two pages placed adjacently, with the longest sides touching, are equal to a larger page with the same height:width ratio, e.g., two pages of A4 equal one of A3. For some applications, the I.S.O. A series is not adequate, and the B series has been introduced to cover a wider range of paper sizes. (Japan has a slightly different set of B series sizes.) The C series is for envelopes.

foolscap

a size of writing or printing paper, now largely obsolete, used for official documents. The size was about 13 $^1/_2 \times 17$ inches (34.25 $\times$ 43 cm), although foolscap folio (half the size), once commonly used in offices, was often referred to simply as "foolscap." The name comes from a watermark picture of a court jester's head originally used on this type of paper.

legal cap

writing paper in tablet form, each sheet measuring 8 $^1/_2$ inches $\times$ 13 to 16 inches (21.5 $\times$ 33 to 40.5 cm), with a ruled margin. Used by attorneys in the U.S. for writing legal documents.

letter

writing paper size commonly used in North and South America, with dimensions 8$^1/_2 \times 11$ inches (21.5 $\times$ 28 cm). It is slightly wider but not as long as its metric counterpart—A4.

atlas

a large size of writing or (more usually) drawing paper generally measuring 26 $\times$ 34 inches (66 $\times$ 86.5 cm).

imperial

a size of writing or drawing paper generally measuring 22 $\times$ 30 inches (56 $\times$ 76.25 cm). This was the largest size of writing paper ordinarily made.

crown

a size of paper measuring 20 $\times$ 15 inches (51 $\times$ 38 cm). Before the U.K. switched to the I.S.O. paper sizes, Metric Crown (quad) paper was introduced, measuring 1.008 $\times$ 768 mm. The book size Metric Crown Quarto measures 189 $\times$ 246 mm (7 $^7/_{16} \times 9$ $^{11}/_{16}$ inches), and Metric Crown Octavo 123 $\times$ 186 mm (4 $^{13}/_{16} \times 7$ $^5/_{16}$ inches). These are slightly different from the old imperial sizes.

demy

a size of writing or drawing paper measuring $22 \, 1/2 \times 17 \, 1/2$ inches (57×44.5 cm). Double Demy measures $35 \times 23 \, 1/2$ inches (89×59.75 cm). Before the U.K. switched to the I.S.O. series, Metric Demy (quad) paper was introduced, measuring 1.128×888 mm. The book size Metric Demy Quarto measures 219×276 mm ($8 \, 5/8 \times 10 \, 7/8$ inches), and Metric Demy Octavo 138×216 mm ($5 \, 7/16 \times 8 \, 1/2$ inches). These differ slightly from the old imperial sizes.

basic size

the usual standard sheet size of paper. The weight in grams of metric paper is based on a basic size of 1 square meter, the area of a piece of A0. U.S. paper is measured in pounds and the size used varies between different grades: for example, bond, envelope, ledger and thin papers have a basic size of 17×22 inches, cover papers 20×26 inches, and book, offset and text papers 25×38 inches. These standard sheet specifications persist, even though in many cases paper is no longer made in such sizes.

basis weight

in North America, the designated fixed weight of a ream of paper (500 sheets), measured in pounds. Also sometimes called the "ream weight" or "substance weight (sub weight)." This was determined using each grade of paper's **basic size**, which can vary considerably between grades. A ream of 20 lb bond paper at its basic size of 17×22 inches (43×56 cm) weighs 20 lb (9 kg). The basis weight of bond paper, which is used in printers and copying machines, typically varies between 13 and 25 lb (6 and 11.5 kg).

G.S.M.

grams per square meter, also (and more correctly) rendered as g/m^2. This measure is used in countries using the I.S.O./A.B.C. series of paper sizes to describe different thicknesses of paper. The basic weight is taken from the A0 size of paper, which is 1 square meter in area. The higher the G.S.M., the thicker the paper. Common photocopier or printer paper is 80 G.S.M. The slightly better-quality 90 and 100 G.S.M. papers are also widely available.

ream

normally 500 sheets of paper, but formerly between 480 and 516 sheets. Paper is usually sold today in reams of 500 sheets, both in the metric (I.S.O./A.B.C.) system and the North American imperial system, although tissue and wrapping paper is sometimes still sold in the U.S. in reams of 480 sheets.

quire

rarely used term denoting 24 or 25 sheets of paper of the same size and quality. A quire of writing paper had 24 sheets, a quire of printing paper 25 sheets. Twenty quires make a ream. The now normal size of a ream at 500 sheets implies a standardization of quire at 25 sheets but in fact neither value of quire has much currency.

folio

originally a large sheet of paper folded once to make two leaves of a book, or a book made up of such sheets. The size of a folio book is very large, something like a "coffee table" book. However, a folio may also simply be a folded sheet of any size. Several folios sewn together make a "signature"; the signatures are then bound together to make the book. Furthermore a sheet of paper in a book numbered on one side only is also referred to as a "folio," and the term is sometimes even used simply to denote a page number.

recto

the right-hand page in an open book. Thus the first page of a book is always a recto page, as are all subsequent odd-numbered pages. From the Latin word *rectus*, meaning "right."

verso

the left-hand page in an open book, the other side of a leaf from a recto page. All even-numbered pages in a book are verso pages. From the Latin word *vertere*, "to turn," because you turn the leaf to see the verso page.

pagination

the numbering of the pages in a book, or the method of such numbering. The word is also used to describe the arrangement and number of pages in a book as shown in a catalog or bibliography.

-mo

a suffix forming nouns referring to the size of a book. This originally related to the amount of times a sheet of paper measuring 19 × 25 inches had been folded, although in practice several different sizes of broadsheet were used. The first three sizes were folio (folded once), quarto or 4to (twice) and octavo or 8vo (three times). Then followed sixteenmo or 16mo (four times), thirty-twomo or 32mo (five times) and sixty-fourmo or 64mo (six times). A folio book had two leaves, or four pages, per sheet whereas a sixty-fourmo book had 64 leaves, or 128 pages, per sheet.

❧ **leading**

the amount of space between lines of type. When type was set by hand—as it was until fairly recently—the typesetter placed strips of lead between the lines of type to create the necessary vertical space between the lines. "Kerning" is the horizontal equivalent in a line of type. Nowadays, of course, type is set electronically, and the size of the leading and kerning can be specified at the touch of a button. The unit of measurement of leading is the **point**. One point is equal to 0.351 mm in the U.S. and Britain, and 0.376 mm in Europe. The leading of this book is set at 12 point.

point

a unit of measurement used in typography for type size, **leading** and other space specifications in a page layout. In the system used in North America and Britain there are approximately 72 points to an inch, 28 to a centimeter, and 12 points make a pica. Type size is defined as being measured from the top of the highest ascender to the bottom of the lowest descender. Different font styles mean that the actual measurement may be different from the size expressed in points. With the advent of personal computers, Adobe Systems launched its PostScript font system which has exactly 72 points to an inch. So, slightly confusingly, there are two systems existing side by side with only a slight difference between them.

pica

a measurement used in typography for the width of columns and other space specifications in a page layout. The etymology of the word "pica" is uncertain, although the medieval Catholic Church's rule-book on daily services was called the Pica, and there is thought to be a link. Twelve points make up a pica, and there are approximately 6 picas to an inch (just under 2 1/2 to a centimeter). The analogous term in the **Didot point** system is the "cicero," also amounting to 12 points.

Didot point

a unit of measurement used in typography throughout Europe to measure type sizes and space specifications. There are approximately 68 Didot points to an inch (each point measures 0.376 mm). The point-based system was originally created in France by Simon Fournier, who used the French type size Cicero as the basis for the system, defining one point as a twelfth of a cicero. In 1770, François Ambroise Didot adapted Fournier's system, redefining a point as 1/72 of a French royal inch (slightly bigger than an imperial inch). The metric system adopted after the Revolution meant the demise of the French inch, but the Didot point system remains.

x-height

the distance between top and bottom of the body of letters, in a line of type—i.e., excluding the ascenders or descenders. It is possible for letters of different typefaces to have the same **point** size but different x-heights.

To achieve an adequate amount of space between lines so that the ascenders and descenders on letters do not collide with the type on the line above or below, the type requires leading of a greater point size. The type in this caption is 7.5 pt, set on 10 pt leading.

textiles and cloth

❧ denier

a former unit for measuring the density of yarn. One denier is the density of a yarn 9,000 meters long weighing 1 gram. The denier was used especially for silk or nylon thread, and was well known as a measure of the sheerness of ladies' stockings. It has been superseded by the metric unit tex (formerly known as drex), expressed as the weight in grams of a yarn 1 kilometer long, and its derived units the decitex (dtex, equal to $1/10$ tex), millitex (mtex, $1/1,000$ tex), and kilotex (ktex, 1000 tex). The density of multi-fold yarns can be accurately expressed in terms of the resultant value; for example, two threads each of 10 tex would result in a yarn of density R20 tex, or 10 tex × 2 (the R meaning "resultant").

Women's hosiery is expressed in denier. Ultra sheer (less than 10 denier) gives a bare-leg appearance. Sheer is 10–20 denier. Semi-opaque, stronger and longer-lasting than sheer, is 25–30 denier; and opaque is 40 denier plus, designed to cover the leg completely.

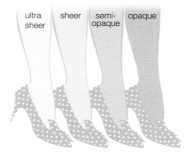

gram per square meter (g/m²; gsm)

a unit of density of fabric, or more commonly, paper and board, expressed in terms of mass and surface area.

thread-count

the number of picks (for length) or ends (for width) per unit length of a fabric. For finer fabrics this is normally expressed as threads per centimeter (or threads per inch in traditional U.K. and U.S. usage), but for coarse cloth as threads per 10 centimeters.

evenweave

woven cloth in which warp and weft threads are of the same thickness and tension, and with the same number of threads per unit length and width.

gauge

a unit of the fineness of knitted fabric. It can either refer to the number of needles per unit length of the needle bed or needle bar of the knitting machine, usually expressed as needles (or loops) per inch or per centimeter. Gauge can also refer to the size of needles used to knit a fabric of that density of knit.

plain weave   basket weave

twill (Z)   twill (S)

Different weaves used in cloth, to give different decorative or textural effects clockwise from top left: plain weave, basket weave, twill 2/2 z-wale and twill 2/2 s-wale.

❧ ply (fold)

when applied to yarn, an indicator that the yarn is made up of more than one thread; e.g., a yarn composed of two threads twisted together would be a two-ply or two-fold yarn. The term is also used to indicate the number of folds or layers of a fabric, paper or board.

skein (lea)

a traditional unit of variable length used for measuring yarn. The value of a skein depends on the type of yarn, and can vary from manufacturer to manufacturer, but generally accepted values include:

wool and worsted	*1 skein = 80 yards (73 m)*
cotton and silk	*1 skein = 120 yards (110 m)*
linen	*1 skein = 300 yards (274 m)*

The skein is sometimes equal to ½ **hank**, especially in the cotton and wool trades. In the U.S. it is more commonly called a lea.

hank

a traditional unit of variable length used for measuring yarn, its value depending on the type of yarn, and varying from one manufacturer to another. For cotton and wool, it is normally equal to seven skeins (or leas), but in the U.S. can mean 1,600 yards (1,463 m) of wool, and local usage gives several different values.

ell

a traditional unit of length used for measuring cloth. The word derives ultimately from the Latin *ulna* (elbow), which is now used as the name for a bone of the forearm, and ells were originally measured using the forearm in one way or another. As such, it was a very variable measurement, and different values evolved in different countries in Europe. In England, the ell came to be equal to 45 inches or 1.25 yards (1.143 m), whereas the Scottish ell was considerably shorter at 37.2 inches (0.945 m)—more like a yard (with which it is often confused). The French equivalent, the *aune*, was equal to 46.77 inches (1.188 m). In Germanic Europe, the related unit the *elle* was generally shorter than even the Scottish ell, being reckoned as two *fuß* (German feet), and had values ranging from about 23 to 30 inches.

bolt

a unit of length used in buying and selling cloth. Because the width of cloth varies widely depending on the type of fabric and the size of loom used in its manufacture, the actual area of a bolt varies accordingly (see **width**), and even the length of a bolt is open to interpretation. Formerly the bolt may have been measured in **ells**, but today the length of a bolt is either 30 yards (27.43 m), 40 yards (36.58 m), or 100 yards (91.44 m). It is therefore

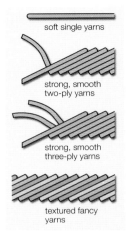

soft single yarns

strong, smooth two-ply yarns

strong, smooth three-ply yarns

textured fancy yarns

The spiral composition of the threads making up a multi-ply yarn is known as the twist, and is measured in turns per unit length (usually per inch or centimeter).

important for buyer and seller to specify which length is meant. and not surprising that the bolt is being replaced by less ambiguous metric measurements.

width

the width of finished cloth is dependent on both the type of yarn and the size of the loom used. and is measured across the warp of the fabric. This varies widely also from manufacturer to manufacturer. but certain traditional measurements are considered standard: for cotton the normal bolt-width is 42 inches (1.067 m). for woolen cloth 60 inches (1.524 m).

absorbency

the capacity of a fabric to absorb liquids. This can be expressed in a number of ways as the result of an absorbency test: e.g.. as the change in weight of a sample of the fabric. the capillary rise of the liquid in a sample in a given time. or the time taken for a given capillary rise.

Shoe sizes remain one area where metrication has not gained a foothold—even the Continental sizing system is somewhat eccentric—and in any case most sizes are only approximate.

shoe sizes

U.K.	U.S. ladies	U.S. men	Continental	Mondopoint
	1			190
	1.5			
	2	1		200
1	2.5	1.5	33	
1.5	3	2	34	210
2	3.5	2.5	34	
2.5	4	3	35	
3	4.5	3.5	35	220
3.5	5	4	36	
4	5.5	4.5	37	230
4.5	6	5	38	
5	6.5	5.5	38	240
5.5	7	6	39	
6	7.5	6.5	39	250
6.5	8	7	40	
7	8.5	7.5	41	
7.5	9	8	42	260
8	9.5	8.5	42	
8.5	10	9	43	270
9	10.5	9.5	43	
9.5	11	10	44	
10	11.5	10.5	44	280
10.5	12	11	45	
11		11.5	46	290
11.5		12	47	
12				300

colorfastness

the capacity of a fabric to retain its original color without fading or running when exposed to sunlight. washing. dry cleaning. rubbing and so on. There are several factors affecting colorfastness. including the type of fiber of the fabric. the type of dye and the treatment used for setting the color.

₰ shoe sizes

a unit describing (usually) the length of shoe in one of several scales in use around the world. Shoe sizes are in any case rather approximate. and depend on the size of last on which the shoe was made. The most commonly used systems are the U.S. and U.K. systems. which are both calibrated in increments of $1/3$ of an inch (8.47 mm). and the Continental or Paris Point system. in increments of 6.66 mm. Mondopoint. a scale in millimeters to specify both length and width of foot. was adopted by the International Standards Organization. but has failed to catch on except in South Africa and parts of eastern Europe.

₰ men's sizes

traditionally. men's clothing sizes have derived from the measurements taken for bespoke tailoring, for example waist, chest, inside leg and sleeve length, in inches in the U.K. and U.S., and in centimeters in Europe. With the advent of mass production, suits and overcoats tended to be sized simply by chest measurement, and shirts simply by collar size, while trousers continue to be sized by both waist and inside leg measurement. Recently the pressures of mass production have led to even more standardized sizing in terms of small (S), medium (M), large (L), and extra large (XL).

men's sizes							
suits/overcoats		shirts (collar size)			socks		
U.K./U.S.	continental	U.K./U.S.	continental		U.K./U.S.	continental	
36	46	12	30–31		9.5	38–39	
38	48	12.5	32		10	39–40	
40	50	13	33		10.5	40–41	
42	52	13.5	34–35		11	41–42	
44	54	14	36		11.5	42–43	
46	56	14.5	37				
		15	38				
		15.5	39–40				
		16	41				
		16.5	42				
		17	43				
		17.5	44–45				

women's sizes							
suits/dresses			brassieres*			hosiery	
U.K./U.S.		continental	U.K./U.S.	continental		U.K./U.S.	continental
8	6	36	32 inches	70 cm		8	0
10	8	38	34 inches	75 cm		8.5	1
12	10	40	36 inches	80 cm		9	2
14	12	42	38 inches	85 cm		9.5	3
16	14	44	40 inches	90 cm		10	4
18	16	46	42 inches	95 cm		10.5	5
20	18	48					
22	20	50					
24	22	52					

*Note: measured across the bust for U.K. and U.S. sizes; under the bust for continental sizes

Buying "off the peg" clothes from abroad or when traveling can be a risky business, as conversion tables often only give approximate equivalents.

ﻮ dress sizes

as with **men's sizes** for clothing, the bust, waist and hip measurements used by traditional dressmakers have given way to a standardized system of sizing women's clothes.

glove sizes

gloves are traditionally sized according to the width measured around the knuckles in inches. There is a slight variation between

S- and Z-twist. The spinning process affects the strength of multi-ply yarns: high-twist yarns are hard and strong, suitable for weaving; low-twist yarns are soft, suitable for knitting.

English and continental sizing, but there is no metric equivalent to this system and it is still widely used, though it is inevitably being superseded by the ubiquitous S, M and L of mass production.

staple length

the average length of the fibers used to spin a yarn. The fibers, known as the staple, can be as short as 3 mm (1/8 inch) in cotton, and as long as 1 meter (about 39 inches) in flax. Man-made fibers are often spun into continuous filament yarn, or used as unspun monofilament, but some are cut or broken into shorter staple lengths of between 5 and 46 cm (2 and 18 inches).

crimp

a measure of the waviness of a fiber. Crimp is expressed as the percentage difference between the length of the fiber in its wavy state and the length when straightened. The crimp of a woven fabric is measured as the percentage difference between the length of the fabric and a yarn taken from it and straightened.

dyeability

the capacity of a fabric to attract and retain dye when put in a dye-bath, also known as substantivity. This is dependent on both the type of fabric and the type of dye; some dyes, known as substantive dyes, can dye cotton and other fibers easily, but where dyeabilty is poor, a compound known as a mordant is applied to the fabric to form a stable complex with the dyestuff.

moisture regain

the weight of water in a fabric compared with its oven-dry weight. Under normal conditions, all dried fabrics regain some moisture from the atmosphere; moisture regain is a measurement of this expressed as a percentage of the dry weight.

twist

the number of turns per unit length of a spun yarn. To hold relatively short staple lengths of fiber in a continuous yarn, they are twisted together in the spinning process. The twist is measured in turns per centimeter (or per inch). The direction of the twist can be seen when the yarn is held vertically; if the turns are diagonal from left to right it is said to have S-twist, if from right to left, Z-twist.

hat sizes

U.K.	U.S.	metric
6⅜	6½	52
6½	6⅝	53
6⅝	6¾	54
6¾	6⅞	55
6⅞	7	56
7	7⅛	57
7⅛	7¼	58
7¼	7⅜	59
7⅜	7½	60
7½	7⅝	61
7⅝	7¾	62
7¾	7⅞	63
7⅞	8	64

As with most clothing sizes, there is a bewildering difference between U.S., U.K. and metric systems. Probably the best advice is: if the cap fits ...

hat sizes

the sizing of hats in North America is based on the circumference of the head in inches divided by pi (equivalent to the diameter if the head were a regular sphere); strangely, the U.K. sizes are equivalent to North American sizes minus 1/8. Happily, the increasingly used metric sizing system is based quite simply on the circumference of the head, although mass production is pushing the use of the even simpler S, M and L.

music

treble

a category of singing voice, normally used to describe the unbroken male voice. The range of the treble voice is roughly the same as the soprano (i.e., from c′ to a″). The word treble is also used adjectivally to describe some instruments with a similar pitch range to the treble voice, such as the treble recorder. It further denotes the most common **clef** sign, used for instruments and voices with the treble pitch range.

♫ soprano

a category of female voice, the highest of the singing voices, with a normal range from c′ to a″. The word soprano is also used adjectivally to refer to instruments with a similar pitch range, such as the soprano saxophone. The diminutive form "sopranino," describes very high instruments such as the sopranino clarinet. A soprano **clef** exists, but is extremely rarely seen.

♫ mezzo-soprano

a category of female singing voice, lying between soprano and contralto, with a normal range from a to f″. A mezzo-soprano clef exists, but is extremely rarely seen.

♫ contralto

a category of singing voice, the lowest of the female voices, with a normal range from g to e″. The term contralto is normally restricted to describing exclusively women's voices within this pitch range.

alto

a category of singing voice, usually male (falsetto, castrato, high tenor or boy) with a normal range from g to c″. Confusingly, contraltos are also frequently referred to as altos. The word alto is also used adjectivally to describe some instruments with a similar pitch range to the alto voice. It further denotes the **clef** used for instruments such as the viola.

counter-tenor

a category of male singing voice, the highest of the adult male's voices, with a normal range similar to the contralto, from g to e″. Until the 19th century, counter-tenor parts were usually sung by a castrato (a singer castrated in boyhood to retain his unbroken voice range), but they are now performed by very high tenors or mature male singers using the supranormal range known as falsetto.

pitch ranges of voices

soprano

mezzo-soprano

contralto

tenor

baritone

bass

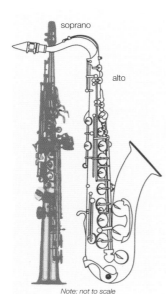

soprano

alto

Note: not to scale

🎵 tenor

a category of male singing voice. with a normal range from c to a'. The word tenor is also used adjectivally to refer to instruments with a similar pitch range. such as the tenor saxophone. It further denotes the **clef** used for instruments such as the tenor trombone. and the modified treble clef used for tenors in much vocal music.

🎵 baritone

a category of male singing voice. lying between bass and tenor. with a normal range from A to f'. A baritone **clef** exists. but is extremely rarely seen.

🎵 bass

a category of male voice. the lowest of the singing voices. with a normal range from F to e'. A singer with a range extending below F is known as a basso profondo (deep bass). The word bass is also used adjectivally to refer to instruments with a similar pitch range. such as the bass clarinet. and to denote the **clef** used for instruments such as the bassoon.

Many instruments are grouped into "families"; the saxophone family, for instance, comprises the soprano, alto, tenor, baritone and bass saxophones.

🎵 interval

the distance between two notes of a diatonic **scale**. described by the number of steps of the scale they include: C up to D is a second. C up to E a third. and so on. An interval also implies a ratio between the frequencies of the two pitches (e.g.. octave = 1:2. perfect fifth = 2:3. perfect fourth = 3:4. major third = 4:5. minor third = 5:6). although these "pure" pitch ratios are modified in **equal temperament** to accommodate 12 exactly equal semitone steps per octave. Non-western and avant-garde music often employ intervals outside the diatonic scales. such as quarter-tones and other microtonal intervals. and cannot be described simply in conventional musical terms.

The intervals shown here are the distance from C to various other notes of the scale; similar intervals can be formed from any note.

tone (whole tone; step)

the interval of a major second, unsurprisingly the sum of two semi-tones. A "pure" tone has the pitch ratio 8:9, but in **equal temperament** this is modified to 1:1.12246204-75 (i.e., the pitch ratio of a semitone to the power of two), so that the octave is divided into six exactly equal whole-tone steps.

semitone (half-step)

the smallest interval in western diatonic and chromatic scales, the distance from one note to its immediate neighbor on the keyboard. A "pure" semitone has the pitch ratio 15:16, but in **equal temperament** this is modified to 1:1.059463094 (i.e. the 12th root of 2), which ensures a division of the octave into 12 exactly equal semitone steps.

octave

an interval in the musical scale with the frequency ratio of 1:2, comprising eight steps of a diatonic scale. Any note and its octave share the same letter name: the octave above A, for example is a, and the octave above c' is c". It is the only interval in **equal temperament** which is "pure," i.e., it is not altered to fit an equal division of the chromatic scale.

equal temperament

a system of tuning which divides the octave into 12 exactly equal intervals. Equal temperament was devised in the 17th century to overcome the problems of tuning instruments, especially keyboard instruments, using the natural harmonic series. Previously, the scale had been based on the exact **pitch** ratios of the intervals of the harmonic series, especially the fifth (a ratio of 3:2) and fourth (2:2). If a scale is built on these intervals the resulting "octaves" above and below the starting point do not correspond exactly to the ratio 2:1, i.e., they begin to go out of tune. A compromise is achieved by "tempering" all the intervals of the scale by very small amounts, so that the pitch ratio of each semitone step is 1:1.059463904, the 12th root of 2.

pitch

the standard which determines the frequency of a given note (usually a', the A above middle C) from which all other notes of the scale can be extrapolated. The standard pitch (often called "concert pitch") of a' = 440 Hz was established by the International Standards Organization (I.S.O.) in 1955, after centuries of confusion when the pitch of a' varied from 400 to 500 Hz. In "authentic" performances, historic pitches such as a' = 430 Hz and a" = 415 Hz are still used.

The term pitch is also used loosely to assign a letter name to a particular note, and thus refers indirectly to its fundamental wave frequency. There are, confusingly, several different systems to specify which octave a particular letter-named note refers to. In the U.S. and Canada, a common usage gives middle C as C_3, the octave above as C_4 and the octave below as C_2, and so on. The convention common in the U.K. and Europe represents middle C as c′, the octave above as c″, the octave above that as c‴, etc. The octave below middle C is given as c, the octave below that as C, and the octave below that as C_1. In both systems pitches between these reference notes are similarly qualified—the tone above middle c′ (C_3) being d′ (D_3), the semitone below being b (B_2).

frequency

the number of vibrations per second of a musical tone, measured in hertz (Hz) or cycles per second. When referring to a complex musical note rich in overtones, the frequency of the fundamental wave determines the pitch.

❧ scale

a sequence of notes, normally spanning an octave, in ascending or descending order of pitch. In western music the most common are the diatonic scales, especially major and minor scales, and the chromatic, pentatonic and whole-tone scales, although many others (often referred to as "modes") have been devised. The diatonic scales include the major and minor scales and the church modes, and consist of seven steps within the octave comprising each of the letter-named notes in order. A sequence of twelve semitone steps is known as a chromatic scale; a sequence of six whole-tone steps forms the whole-tone scale; and any scale of five steps within the octave is known as a pentatonic scale. In non-western music, scales are often formed using pitches outside the chromatic scale involving microtonal intervals.

key

the diatonic scale which forms the basic tonality of a piece of music. A piece of music in a particular key uses mainly the notes of that scale and harmonies derived from those notes. The name of

All major scales follow the pattern: tone, tone, semitone, tone, tone, tone, semitone. Similarly, all other scales follow the patterns as shown below, and can start on any note.

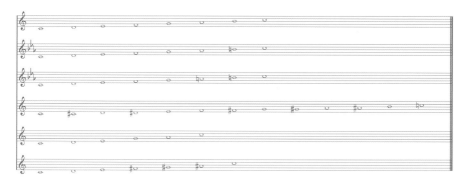

the key is taken from the first note, the tonic, of that scale, and though the music may move to related keys by a process of modulation, it normally starts and ends in that "home" key. Thus a piece in the key of C major would start and end with a chord of C major, and the melody and harmonies would use the notes of the scale of C major as their basis. The concept of key is not, however, universal: western music before the 16th century, non-western music, and much 20th-century avant garde music is not based on diatonic scales, so cannot be described in terms of key or tonality.

clef

a sign placed on the five lines of the musical staff to indicate the pitches of the various lines and spaces. Originally stylized versions of the letter names of the notes they denote—F, G and C—the three styles of clef in use today are, respectively:

𝄢 *F or bass clef*

The F clef, now more commonly called the bass clef, indicates that the fourth line up represents the pitch f, and the other pitches can be deduced from this reference point.

𝄞 *G or treble clef*

Similarly, the G clef (now usually known as the treble clef) assigns the pitch g′ to the second line up on the staff.

𝄡 *C or alto or tenor clef*

The C clef, by contrast, can be placed in several different positions on the staff, but however placed its center determines the line assigned to the pitch c′: when it falls on the center line of the staff the clef is known as the alto clef, when on the fourth line up it is known as the tenor clef. Very rarely it is centered on other lines to become soprano, mezzo-soprano or baritone clefs.

time signature

a sign placed on the staff indicating the meter of the music, i.e., how many beats to the bar (or measure), and what kind of beat. The time signature consists of two figures, one above the other: the lower denotes the unit of measurement relative to the **whole note** (semibreve), the upper indicates how many of these units occur in the measure. Thus a time signature of 3/4 indicates that there are three **quarter-note** (crotchet) beats to the bar, 4/2 that there are four **half-note** (minim) beats to the bar. If the upper figure is a multiple of three, as in 6/8 or 12/8 time, the music is almost always in "compound time"—the units of the lower figure are grouped in threes. So, the time signature 6/8 indicates that there are six **1/8th note** (quaver) units in the bar, but that there are two dotted **quarter-note** (crotchet) beats; similarly 12/8 groups the twelve **1/8th note** units into four dotted **quarter-note** beats.

tempo

the speed at which a piece of music is performed. Until the invention of the **metronome**, indications of tempo in musical scores were normally in the form of verbal instructions, by convention (but not exclusively) in Italian, and were subjective and relative terms giving as much an indication of the mood of a piece as its tempo. The most common of these tempo indications, in descending order of speed are:

prestissimo	extremely fast	*presto*	very fast
allegro	fast	*vivace*	lively
allegretto	quite fast	*moderato*	moderately
andante	at an easy pace	*largo*	broadly, slowish
adagio	quite slowly	*lento*	slowly
grave	seriously, slowly		

The markings may be qualified by terms such as *poco* ("a little" or "rather"), *molto* ("much" or "very"), or *ma non troppo* ("but not too much"). Changes in tempo are indicated by the markings *rit.* (ritenuto or ritardando) or *rall.* (rallentando), meaning becoming slower, and *accel.* (accelerando), meaning becoming faster. Metronome markings are a more accurate means of indicating tempi, but lack the expressive connotations of traditional verbal instructions.

metronome

a mechanical or electrical device for determining tempo. A clockwork metronome was patented by J.N. Maelzel in 1815 (though it is possible it was invented by one Winkel some years earlier), and in the late 20th century several electronic devices appeared. In essence, a metronome provides an audible (and sometimes visible) regular beat at a specific frequency per minute. This establishes an appropriate tempo for a given piece of music. With the invention of Maelzel's metronome, composers began to assign "M.M." numbers to their work, giving the number of beats per minute in the form M.M. = 120 (a tempo of 120 beats per minute). Later, the unit of the beat was specified in the metronome marking, for example q = 120 (a tempo of 120 crotchet beats per minute), a form which has persisted to the present.

double note (double whole note; breve)

the longest note value in current use in western music, although its appearance is becoming rare. Its duration is equivalent to two **whole notes**, four **half-notes**, or eight **quarter-notes**, and its actual duration in time is dependent on the tempo. Paradoxically, the word breve comes from the Latin meaning "short"; originally the breve was the short note of medieval musical notation, but the longer notes of this system have fallen out of use, and shorter note values have been devised to subdivide the breve.

whole note (note: semibreve)

the longest note value in common usage in contemporary music. As its name implies. its duration is equivalent to half a **double note**. two **half-notes** or four **quarter-notes**.

half-note (minim)

a note value whose duration is. as its name implies. half that of a **whole note** (semibreve). or. following the logical sequence. the equivalent of two **quarter-notes**.

quarter-note (crotchet)

a note value whose duration is. as its name implies. quarter that of a **whole note** (semibreve). In the most commonly used time signatures—for example. 4/4. 3/4 and 2/4—the measure (or bar) is divided into quarter-note beats.

1/8th note (quaver)

a note value equivalent to half a **quarter-note**. or. as its name implies. to 1/8 of a **whole note**.

1/16th note (semiquaver)

a note value equivalent to half an **1/8th note**. or. as its name implies. to 1/16 of a **whole note**.

1/32nd note (demisemiquaver)

a note value equivalent to half a **1/16th note**. or. as its name implies. to 1/32 of a **whole note**.

1/64th note (hemidemisemiquaver)

a note value equivalent to half a **1/32nd note**. or. as its name implies. 1/64 of a **whole note**.

dotted notes

notes with a dot placed immediately after them. which increases their length by a half. A **half-note** (minim). which has a time value of two **quarter-notes** (crotchets). when dotted becomes the equivalent of three **quarter-notes**: and a dotted **quarter-note** has the time value of three **1/8th notes** (quavers). The length of a note can be further increased by the placement of an additional dot immediately after the first dot. These double-dotted notes follow a similar logic, with the second dot increasing the time value by a further quarter of the original note: a **half-note** when marked with a double dot becomes the equivalent of a **half-note** plus a **quarter-note** plus an **1/8th note**.

By using dotted notes. values between those shown below can be created. A rest can also be dotted to increase its value by a half. but this is more commonly done by adding a rest of the next value down.

pianissimo

a dynamic marking (i.e., indication of loudness) in music, from the Italian meaning "very soft." It is usually abbreviated to pp in musical scores, and like all such dynamic markings is a relative and subjective, rather than absolute, measurement of loudness; the range of dynamics from pianissimo through to fortissimo does, however, give a remarkable accuracy of nuance in practice, especially when combined with instructions such as *cresc.* (crescendo—getting louder) and *dim.* (diminuendo—getting softer). Although pianissimo is normally considered the quietest of the dynamic markings, ppp (very, very soft) is not uncommon. There are occasional instances of $pppp$ (very, very, very soft), and even more extreme dynamic markings exist.

piano

a dynamic marking in music, from the Italian meaning "soft." usually abbreviated to p in musical scores.

mezzopiano

a dynamic marking in music, from the Italian meaning "medium soft" (somewhere between piano and mezzoforte), usually abbreviated to mp in musical scores.

mezzoforte

a dynamic marking in music, from the Italian meaning "medium loud" (not as loud as forte, but louder than mezzopiano), usually abbreviated to mf in musical scores.

forte

a dynamic marking in music, from the Italian meaning "loud." usually abbreviated to f in musical scores.

fortissimo

a dynamic marking in music, from the Italian meaning "very loud." usually abbreviated to ff in musical scores. Occasionally composers will demand something even louder, indicating this with fff or $ffff$, or even further fs to indicate extreme loudness.

some familiar sounds, measured in decibels	
10 dB	rustle of leaves
20 dB	library noise
30 dB	soft whisper
50 dB	light traffic noise
70 dB	vacuum cleaner/train
100 dB	thunder
130 dB	jet aircraft at take-off
180 dB	space rocket at take-off

✌ decibel (dB)

a unit of intensity of sound, derived from the bel, but in much more common usage. Although it is a far more accurate measure of the loudness of a sound than the dynamic markings of musical notation, its use is still restricted to science rather than music. The decibel is a logarithmic unit, each 10 dB step representing a tenfold increase in sound intensity, or a doubling of the perceived loudness, on a scale starting with 0 dB as the threshold of human hearing.

bel

a unit of sound intensity, equivalent to 10 decibels. Named after Alexander Graham Bell (1847–1922).

photography

A.S.A. speed

the American Standards Association (A.S.A.) system of film speeds, now adopted by the I.S.O. (International Standards Organization). It is an arithmetic system, so 200 A.S.A. is twice as fast as 100, 400 is twice as fast as 200, etc. The faster the speed, the more "grainy" the resulting photographs tend to be, although this is less significant now because of considerable improvements in film quality over the past two decades. The "speed" is actually a measure of how much light the film needs to record a photographic image. Indoor photography without a flash needs a higher speed film, and faster film is also appropriate for "action" shots.

D.I.N. speed

a logarithmic rather than arithmetic system of film speeds (see **A.S.A. speed**), developed in Germany. Each increase represents 1/3 **stop**. Film boxes often show two values: the I.S.O. film speed followed by the D.I.N. speed, e.g., 100/21. Strictly speaking, this is the recommended "exposure index" and denotes the setting on the camera. The actual film speed may be slower.

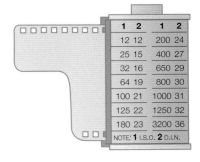

1	2	1	2
12	12	200	24
25	15	400	27
32	16	650	29
64	19	800	30
100	21	1000	31
125	22	1250	32
180	23	3200	36

NOTE: **1** I.S.O. **2** D.I.N.

Ilford speed

an alphabetical system of film speeds, once used by the Ilford film company. "A" was equivalent to A.S.A. 4, "B" A.S.A. 8, "C" A.S.A. 16, and so on, up to "H" which was A.S.A. 512.

For amateur photographers, a film speed of 200 I.S.O. is usually satisfactory.

time lapse

several photographs of the same subject taken at intervals from the same position, showing changes to the state of the subject, e.g., the different stages in the opening of a flower bud. The photographs can be taken with several seconds, minutes, hours or even days between them, and then joined together to make a film record of the event, giving the impression that it is occurring very quickly.

lag time

the time between pressing the shutter control and the actual taking of a photograph. On a conventional (film) camera, this is usually negligible, but early digital cameras had a significant total lag time, meaning that the camera had to be kept pointing at the subject for a few seconds after the shutter release was depressed. This was a major problem with action shots, as the action could have finished when the photograph was actually taken. The latest, high-specification digital cameras have a considerably reduced lag time, and react in the same way as conventional cameras.

shutter speed

the time the camera's shutter is held open to allow light to reach the film. It is extremely short in most cases, and the setting depends on film sensitivity, aperture setting, the amount of daylight, and whether flash is used. A typical shutter speed in normal daylight is $^1/_{125}$ second. Very fast shutter speeds are needed for action shots and slow speeds for night photography or special effects.

f-number

the ratio of the equivalent **focal length** to the diameter of the camera's aperture. An f number of f/1.7 means that the focal length of the lens is 1.7 times as great as the diameter. A camera with an 80-mm lens set at f/16 would mean an aperture opening of 5 mm. The lower the f-number the greater the amount of light that reaches the film.

stop

on the barrel of a camera lens, a fixed position for an **f-number** setting. Rotating the control takes it through several steps, each of which relates to a certain f-number: f/22, f/16, f/11, f/8, f/5.6, f/4, etc., each signifying a change in the aperture setting. The term "stop" is also used to measure the **dynamic range** of photographs.

The amount of light passed through a lens is inversely proportional to the square of the f number. Modern lenses use a standard f-stop scale with the numbers in the series increasing by a factor of root 2 at each step: f/1, f/1.4, f/2, f/2.8, f/4, f/5.6, f/8, f/11, f/16, f/22, f/32, f/45 and f/64. The values of the ratios are rounded off to make them simple to write down.

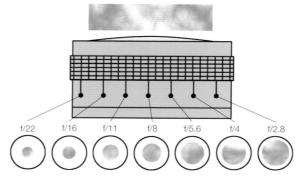

f/22   f/16   f/11   f/8   f/5.6   f/4   f/2.8

exposure value (E.V.)

a value given to a combination of **shutter speed**, film speed and **f-number**. The E.V. is 0 when there is a combination of an aperture setting of f/1 with a shutter speed of 1 second at a film speed of 100 I.S.O. Each time the amount of light is halved, the E.V. increases by one. So with the same aperture setting and film speed, halving the shutter speed will increase the E.V. to 1.

gamma

a measure of contrast in a photograph relating to midtones. It is of particular interest in adjusting the contrast of digital photographs. Adjusting the contrast using photo-editing software often produces intense brightness and darkness; gamma correction is more subtle.

dynamic range

the range of light intensity levels in a photograph, from the darkest shadows to the brightest areas. Some photographic media have better dynamic ranges than others, and the same is true for digital cameras. It is measured in stops, each stop representing a doubling of the intensity of light.

❧ color temperature

a measure of the color of light produced by a light source. The "temperature" of the light source is the temperature at which a theoretical object called a "black body" would produce the same mixture of wavelengths of light. (A black body absorbs all the light and other electromagnetic radiation that strikes it, and emits a mixture of wavelengths that vary with temperature.) Normally expressed in **kelvins**, color temperature for natural light typically ranges from 3,000 K (sunrise or sunset) to about 10,000 K for a heavily overcast sky. Normal daylight is around 5,000 K. For comparison, the color temperature of a candle is only 1,200 K, and that of ordinary household lightbulbs a little below 3,000 K—although specialist lighting can produce almost any temperature desired.

temperature (K)	color temperature
	light source
9,000–12,000 K	blue sky
6,500–7,500	overcast sky
5,500–5,600 K	electronic photo flash
5,500 K	sunny daylight around noon
5,000–4,500 K	Xenon lamp/light arc
3,400 K	1 hour from dusk/dawn
3,400 K	tungsten lamp
3,200 K	sunrise/sunset
3,000 K	200W incandescent lamp
2,680 K	40W incandescent lamp
1,500 K	candlelight

Examples of color temperature. Paradoxically, the color tones that we consider "warmer"— yellows, oranges and reds—have a lower color temperature than the "colder" colors—blue through white. Candlelight and indoor electric light have much lower color temperatures than daylight. Color film is designed for daylight, and so often does not give an accurate rendering of indoor light. Tungsten film is available specifically for indoor applications.

mired ("micro reciprocal degrees")

a measure of **color temperature**, occasionally also used for very high physical temperatures. It is the number produced when the reciprocal of the temperature in **kelvins** is multiplied by one million. For example 40,000 K has the reciprocal 25×10^{-6}, making it equal to 25 mired.

light meter

a device to measure the intensity of light from a subject, allowing camera settings to be adjusted so that the film receives the correct exposure. Also known as an exposure meter.

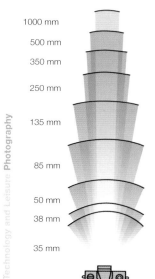

1000 mm
500 mm
350 mm
250 mm
135 mm
85 mm
50 mm
38 mm
35 mm

212

Lenses with a focal length of 50 mm or 55 mm are described as normal, because they replicate a scene as it appears with the naked eye. Wide-angle lenses have a shorter focal length, and telephoto lenses have a longer focal length.

focal length

the distance from the center of the camera's lens to the point of focus.

subject distance

the distance of the subject of a photograph from the lens of the camera. In practice, the focus can be on a spread of distance, depending on the depth of field. The depth of field means the distance between the nearest and farthest objects that appear in sharp focus. The farther the distance of the subject from the camera, the greater the depth of field around it. So a subject far away from the camera is easier to keep in focus than one that is close. Perspective can be altered by shortening or lengthening the subject distance (not by altering the **focal length**, as is sometimes stated).

macro

photography of small objects, either showing them actual size or magnifying them so that they appear a lot bigger than they really are. The normal definition is a range from 1:1 to 10:1. By this method we can "see" objects in much greater detail than with the naked eye, which cannot focus when very close to an object. The enlargement required can be achieved with bellows or extension tubes, close-up supplementary lenses or macro lenses. The definition of macro photography is often extended to include photography with microscopes to show creatures that are not visible to the naked eye at all.

telephoto

a telephoto lens has a number of elements whose effect is to make a subject appear closer to the camera. Telephoto lenses have a focal length longer than the normal 50 mm or 55 mm. They are useful for wildlife or portrait photography, allowing close-ups without intimidating the subject. Their disadvantages are that they reduce the depth of field and that they increase the risk of camera shake.

picture angle

the angle of the coverage of a lens. It is the angle formed at the center of the lens where a line drawn backward from the top left corner of the frame meets a line drawn from the bottom right corner. The longer the focal length, the narrower the picture angle.

back focus

a problem with focusing, when the actual focus is on a point behind the subject. It can mean that the subject is blurred and is often the result of failing to lock autofocus on the subject before the shot is taken. "Back focus" can also mean the distance between the back of the lens and the focal plane.

zoom (x x)

a zoom lens has a variable **focal length** and yet is able to keep the subject in focus at any level of magnification. It is therefore different from a simple **telephoto** lens. Dividing the maximum focal length by the minimum gives the optical zoom or magnification factor. Thus a zoom lens that varies between 35 mm and 280 mm has an optical zoom or magnification factor of 8x.

zoom (mm)

zoom lenses have different ranges of **focal length**. A lens described as 35–280 has a focal length ranging from 35 mm to 280 mm.

digital zoom

most digital cameras have a digital zoom facility, even if they do not have optical zoom. Digital zoom works by enlarging the pixels in an image. The more the pixels are enlarged, the more detail is lost. Also, the picture is cropped, leaving only the central portion.

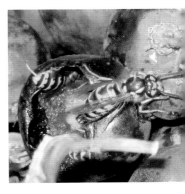

Macro photography can record images in detail not seen with the naked eye.

megapixel

a million pixels. The resolution of pictures taken with digital cameras is measured in megapixels. The higher the number of megapixels, the higher the definition in the photograph produced, so the greater the enlargement possible. Three megapixels will give a resolution of $2,048 \times 1,536$ pixels, for which a common print size at 300 dpi is about 5×7 inches (12.5 by 18 cm). Digital cameras at the lower end of the market may have a maximum resolution of 1.3 megapixels, those at the upper end can reach 12 megapixels.

effective pixels

the pixels that actually record images in a digital camera. The number of effective pixels is more important than the total of actual pixels, because the effective pixels govern the image resolution (some of the pixels are not used to record picture information). Digital cameras may be advertised as offering an "interpolated" image with nearly twice as many pixels as its effective pixels. This is a software enhancement that spreads the pixels out and fills in the gaps with pixels of appropriate colors. It is less satisfactory than having a camera with that amount of effective pixels, but is a cheaper alternative.

35 mm

the most common size (width) of film, it also denotes cameras designed to take this film. The 35 mm size is felt generally adequate to provide good-quality photographs, though professional photographers may use cameras and film twice as wide. Some 35 mm cameras are of the single lens reflex variety, with a large amount of adjustment possible; others are the "point and shoot" type, which focus automatically.

There are, typically, three ways of writing a (finite) large number: it can be written out in full (for example, 1,000,000), written using scientific notation (1×10^6), or it can be given a name (one million). The first two possibilities are nicely unambiguous (providing you assume—or indicate—that the number is in base ten, of course), but the same names can be used for different numbers, causing problems and misunderstanding. The only exceptions to this are the Googol and Googolplex (defined in the "mathematics" section), both of which are too large to be of any actual use.

Two different naming schemes are used in different parts of the world (in addition to the obvious differences of language), both based on the use of Latin prefixes (bi-, tri-, etc.) with "-illion." To further confuse the issue, many countries have used both at different times. The "short" (also sometimes called the "American") scale is based on the number of times a thousand has been multiplied by itself, so a billion is a thousand thousand thousand (1,000,000,000). The "long" system uses the number of times that one has been multiplied by a million, making one billion equal a million million—a thousand times the "short" billion. This used to be widely referred to as the "British" system, but the British government (followed by the rest of the country) officially stopped using the scale in 1974, even though most people still understand both systems.

It is perhaps also worth noting that the "American" system was actually introduced to North America by the French during the 18th century—though the French themselves decided in 1948 to return to the "long" system.

In recent years some people, prominently including Russ Rowlett, Clinical Professor of Education and Adjunct Professor of Mathematics at the University of North Carolina, have suggested scrapping both existing systems and moving to an entirely new one. The logic for this is that since neither the countries of North America nor those of Europe are likely to voluntarily abandon their preferred naming system for that used by the other, the current confusion is likely to continue indefinitely. The proposed solution is to replace both with a new system of names based on prefixes in Greek (similar to those of the SI) rather than Latin, thus requiring both sides of the argument to change—while not resulting in either being seen to have "given way" to the other. There is, however, little support for the plan, since no one wishes to give up their current usage.

Mathematicians—and philosophers—also find themselves needing to distinguish between different infinite numbers. The infinity that is the number of integers is called "aleph-null," $\aleph^0$ (using the first letter of the Hebrew alphabet), as is any other infinity of a "similar" size. The number of irrational numbers (of which there are an $\aleph^0$ infinity) between two consecutive integers is equal to $\aleph^0$ multiplied by itself—clearly a much larger (infinite) number, called $\aleph^1$. $\aleph^1$ multiplied by itself is $\aleph^2$, and so on. While the distinction between these is useful for some mathematical purposes, however, it is unlikely ever to affect everyday life.

scientific notation	American name	European name	"Rowlett name"
10^3	thousand	thousand	thousand
10^6	million	million	million
10^9	billion	milliard	gillion
10^{12}	trillion	billion	tetrillion
10^{15}	quadrillion	billiard	pentillion
10^{18}	quintillion	trillion	hexillion
10^{21}	sextillion	trilliard	heptillion
10^{24}	septillion	quadrillion	oktillion
10^{27}	octillion	quadrilliard	ennillion
10^{100}	Googol	Googol	Googol
10^{googol}	Googolplex	Googolplex	Googolplex

appendix two: **units of the SI (Système Internationale)**

base and derived units

SI base units

Length	meter (m)
Mass	kilogram (kg)
Time	second (s)
Electric current	ampere (A)
Thermodynamic temperature	kelvin (K)
Luminous intensity	candela (cd)
Amount of substance	mole (mol)

Supplementary units

plane angle	radian (rad)
solid angle	steradian (sr)

Derived units

area	square meter
volume	cubic meter
velocity	meter per second
angular velocity	radian per second
acceleration	meter per second squared
angular acceleration	radian per second squared
frequency	hertz (Hz)
rotational frequency	reciprocal second
density and concentration	kilogram per cubic meter
momentum	kilogram meter per second
angular momentum	kilogram meter squared per second
moment of inertia	kilogram meter squared
force	newton (N)
moment of force, torque	newton meter
pressure and stress	pascal (Pa)
dynamic viscosity	pascal second
kinematic viscosity	meter squared per second
surface tension	newton per meter
energy, work, and quantity of heat	joule (J)
power and radiant flux	watt (W)
temperature	degree Celsius (°C)
thermal coefficient of linear expansion	reciprocal kelvin
heat flux density and irradiance	watt per square meter
thermal conductivity	watt per meter kelvin
coefficient of heat transfer	watt per square meter kelvin
heat capacity	joule per kelvin
specific heat capacity	joule per kilogram kelvin

base and derived units

Derived units continued

entropy	joule per kelvin
specific entropy	joule per kilogram kelvin
specific energy and specific latent heat	joule per kilogram
quantity of electricity, electric charge	coulomb (C)
electric potential, potential difference, electromotive force	volt (V)
electric field strength	volt per meter
electric resistance	ohm (Ω)
electric conductance	siemens (S)
electric capacitance	farad (F)
magnetic flux	weber (Wb)
inductance	henry (H)
magnetic flux density, magnetic induction	tesla (T)
magnetic field strength	ampere per meter
luminous flux	lumen (lm)
luminance	candela per square meter
illuminance	lux (lx)
radioactivity	becquerel (Bq)
radiation absorbed dose	gray (Gy)

SI prefixes:
multiples and submultiples

yocto	y	0.000 000 000 000 000 000 000 001
zepto	z	0.000 000 000 000 000 000 001
atto	a	0.000 000 000 000 000 001
femto	f	0.000 000 000 000 001
pico	p	0.000 000 000 001
nano	n	0.000 000 001
micro	μ	0.000 001
milli	m	0.001
centi	c	0.01
deci	d	0.1
deca	da	10
hecto	h	100
kilo	k	1 000
mega	M	1 000 000
giga	G	1 000 000 000
tera	T	1 000 000 000 000
peta	P	1 000 000 000 000 000
exa	E	1 000 000 000 000 000 000
zetta	Z	1 000 000 000 000 000 000 000
yotta	Y	1 000 000 000 000 000 000 000 000

derived units	
Length	
picometer	pm
ångström	Å
nanometer	nm
micrometer (micron)	μm
millimeter	mm
centimeter	cm
decimeter	dc
meter	m
hectometer	hm
kilometer	km
megameter	Mm
international nautical mile	n mile (= 1 852 m)

Area	
square millimeter	mm^2
square centimeter	cm^2
square decimeter	dm^2
square meter	m^2
are	a (= 100 m^2)
decare	daa
hectare	ha
square kilometer	km^2

Volume and capacity	
cubic millimeter	mm^3
cubic centimeter	cm^3 (cc)
cubic decimeter	dm^2
cubic meter	m^3
cubic decameter	dam^3
cubic hectometer	hm^3
cubic kilometer	km^3
microliter (lambda)	μl
milliliter	ml
centiliter	cl
deciliter	dl
liter	l (L)
hectoliter	hl
kiloliter	kl

derived units	
Mass (weight)	
nanogram	ng
microgram	μg (mcg)
milligram	mg
metric carat	CM (= 200 mg)
gram	g
mounce (metric ounce)	
hectogram	hg
glug	kgm (= 0.980665 kg)
kilogram	kg
metric technical unit of mass (metric slug)	(= 9.80665 kg)
quintal	q (= 100 kg)
megagram	Mg
tonne (millier)	t

Force	
micronewton	μN
dyne	dyn (= 10 μN)
millinewton	mN
pond	p (= 9.80665 mN)
centinewton	cN
crinal	(= 10 000 dyn)
newton	N
kilogram-force (kilopond)	kgf (kp) (= 9.80665 N)
kilonewton (sten, sthène)	kN
meganewton	MN

Pressure and stress	
micropascal	μPa
millipascal	mPa
microbar (barye)	μbar
pascal	Pa
millibar (vac)	mbar (mb)
torr	(= ca. 133.322 Pa)
kilopascal (pièze)	kPa
technical atmosphere	at (= 9.80665 Pa)
bar	bar (b)
standard atmosphere	atm (= 101 325 Pa)
megapascal	MPa
hectobar	hbar
kilobar	kbar
gigapascal	GPa

derived units	
Dynamic viscosity	
centipoise	cP
poise	P (= 100 mPa s)
millipascal second	mPa s
pascal second	Pa s
Kinematic viscosity	
centistokes	cSt
stokes	St (= cm^2/s)
Energy, work and quantity of heat	
erg	(= 10^{-7} J)
millijoule	mJ
joule	J
kilojoule	kJ
megajoule	MJ
kilowatt hour	kWh
gigajoule	GJ
terajoule	TJ
Power	
microwatt	μW
milliwatt	mW
watt	W
kilowatt	kW
megawatt	MW
gigawatt	GW
terawatt	TW
metric horsepower	ch (cv, CV, PS or pk) (= 735.498 W)
Temperature	
degree Celsius	°C
kelvin	K
Electricity and magnetism	
picoampere	pA
nanoampere	nA
microampere	μA
milliampere	mA
ampere	A
kiloampere	kA
picocoulomb	pC
nanocoulomb	nC
microcoulomb	μC
millicoulomb	mC

derived units	
Electricity and magnetism continued	
coulomb	C
kilocoulomb	kC
megacoulomb	MC
microvolt	μV
millivolt	mV
volt	V
kilovolt	kV
megavolt	MV
microhm	μΩ
milliohm	mΩ
ohm	Ω
kilohm	kΩ
megohm	MΩ
gigohm	GΩ
microsiemens	μS
millisiemens	mS
siemens	S
kilosiemens	kS
picofarad (puff)	pF
microfarad	μF
farad	F
weber	Wb
picohenry	pH
nanohenry	nH
microhenry	μH
millihenry	mH
henry	H
nanotesla	nT
microtesla	μT
millitesla	mT
tesla	T
Luminous flux	
lumen	lm
lux	lx (= lm/m^2)
Radiation	
becquerel	Bq
kilobecquerel	kBq
megabecquerel	MBq
gigabecquerel	GBq
gray	Gy (= J/kg)

a, A acceleration (length/time^2), atomic mass (total no. of protons and neutrons in an atom)

A Amperes* (electric current = C/s), Angstroms* (length = 10^{-10} m), amplitude (length)

b intercept of a linear graph, drag coefficient (mass/time)

B magnetic field (force/current)

c speed of light (2.998×10^8 m/s), specific heat (energy/mass × temp.), concentration (number/volume), speed of sound

cal calories* (energy = 4.186 J)

cc cubic centimeter

c g group velocity

c p phase velocity

C Celsius* (temp.), coulombs* (electric charge), capacitance (charge/electric potential), heat capacity (energy/temp.), concentration

Cal kilocalories* (energy)

Ci Curie* (unit of radiation = to 3.7×10^{10} decays/s

d distance

D diffusion constant (area/time)

db decibels (relative intensity)

e electron, charge of an electron (1.602×10^{-19} C)

eV electron Volts* (energy = 1.602×10^{-19} J)

E energy (force × length, mass × velocity^2), electric field (force/charge)

f frequency (1/time), focal length

f, F force (mass × acceleration)

F flow (volume/time), Farads* (capacitance = C/V), Fermi* (length = 10^{-15} m)

g grams* (mass), acceleration due to gravity (9.81 m/s^2); sometimes centrifugal acceleration

G Newton's constant (6.673×10^{-11} N m^2/kg^2), Gauss* (magnetic field = 10^{-4} T), free energy

h height, Planck's constant (angular momentum = 6.626×10^{-34} J s), latent heat (energy/mass)

hr hours* (time = 3600 s)

H enthalpy (energy)

Hz hertz (1/s)

I moment of inertia (mass × length^2), current (charge/time), intensity (power/area), image distance

J Joules* (energy = N m), flux (number/area time)

k Boltzmann constant (1.381×10^{-23} J/K), spring constant (force/length), thermal conductivity (power/length temp.), wave number (1/length) (1/length)

K kinetic energy, kelvin* (temp. = C + 273.15)

l length, liters* (volume = 1000 cc), orbital quantum number (dimensionless, denotes angular momentum), mean free path (length)

lb pounds* (weight: 1 kg weighs 2.2 lb)

L angular momentum (momentum × length, moment of inertia × angular velocity)

m mass, meters* (length), slope of a linear graph, magnetic moment (current × area), magnetic quantum number (dimensionless, denotes orientation of angular momentum)

me mass of electron (9.109×10^{-31} kg)

mn mass of neutron (1.675×10^{-27} kg)

mp mass of proton (1.673×10^{-27} kg)

mi miles* (length = 1.61 km)

min minutes* (time = 60 s)

mmHg millimeters mercury (pressure = 1333 dynes/cm^2)

M molecular weight (mass/mol), magnification (dimensionless)

n numbers of mols (dimensionless), no. of loops (dimensionless), neutron, principal quantum number (dimensionless, denotes energy level), index of refraction (dimensionless)

N Newtons* (force = kg m/s^2), no. of particles, neutron number (no. of neutrons in an atom)

N A Avogadro's number (dimensionless no. of objects in a mol = 6.022×10^{23})

O object distance

p proton

P momentum (mass × velocity), pressure (force/area), power (energy/time)

Pa Pascals* (pressure = N/m^2)

q, Q charge

Q heat (energy)

r radius (length), distance, rate (velocity)

R resistance (potential/current), gas constant (8.31 J/mol K)

Re Reynolds number (dimensionless)

s seconds*, sedimentation coefficient (time), spin quantum number (dimensionless), lens strength (1/length)

S entropy (energy/temp.)

t time

T Tesla* (magnetic field = N/A m), temp.

U potential energy (mechanical, elastic, electrical), internal energy

v velocity (length/time), specific volume (volume/mass)

V velocity, volume (length^3), electric potential (electric field × length), voltage, Volts* (N m/C)

W Watts* (power = J/s), weight (force), work (energy)

x, X horizontal position

y, Y vertical position

z, Z vertical position in 3D problems, atomic number (no. of protons in an atom), valence

α **alpha** angular acceleration (radians/time^2), Helium nucleus (2 p + 2 n)

β **beta** electron

Δ **delta** finite change

δ **"d"** instantaneous rate of change

ε **epsilon** electrical permittivity (e0 = 8.854×10^{-12} F/m), emissivity (dimensionless), efficiency (dimensionless)

ϕ **phi** angle

γ **gamma** electromagnetic radiation, photon

η **eta** viscosity (Poise = dyne × s/cm^2 = g/cm × s)

κ **kappa** dielectric coefficient (dimensionless)

λ **lambda** wavelength

μ **mu** magnetic permeability (m0 = $4 p \times 10^{-7}$ T m/A)

ν **nu** frequency or rate of revolution (1/time)

θ **theta** angle, angular position

ρ **rho** density (mass/volume), resistivity (resistance × length)

σ **sigma** Stefan-Boltzmann constant (5.67×10^{-8} W/m^2 K+)

Σ **sigma** summation

τ **tau** torque (force × length, moment of inertia × angular acceleration), radioactive half-life

ω **omega** angular velocity or angular frequency (radians/time)

Ω **omega** Ohms* (resistance = volt/ampere)

NOTE: Asterisk denotes units.
Dimensions and values of constants shown in parentheses.

index

Index

acknowledgments

The publishers wish to acknowledge the
following for their photographic images:

David Evans (p. 213); Nova Development Corporation
(Art Explosion) (pp. 6, 8, 13, 24, 27, 29, 66, 117);
Rebecca Saraceno (p. 65)